RELIABILITY MODELLING

A Statistical Approach

Linda C. Wolstenholme

Senior Lecturer
City University, London

CHAPMAN & HALL/CRC

Boca Raton   London   New York   Washington, D.C.

Library of Congress Cataloging-in-Publication Data

Wolstenholme, Linda C.
  Reliability modelling : a statistical approach / Linda C.
Wolstenholme.
    p. cm.
  ISBN 1-58488-014-7 (alk. paper) ✓
  1. Reliability (Engineering)--Statistical methods. I. Title.
TA169.W65 1999
620′.00452—dc21   99-19565
                                                                                        CIP

To the Memory of my Mother and Father

Contents

List of figures

Preface

The motivation for writing this book is a common one. I have spent much time teaching reliability to master's degree students, who come principally from an engineering background, and it has always been difficult to find a single text as the main reading for the course. However, this book is not aimed solely at postgraduates. It is equally suitable for undergraduates studying reliability as part of a mathematics, computing, engineering or systems degree, for researchers needing to consider the reliability implications of their work, and for engineers of all disciplines who have a practical interest in reliability.

The reader is assumed to have a level of mathematics equivalent, say, to UK 'A' level. Many graduates and postgraduates with a higher qualification than this often find that mathematical skills suffer through lack of use, so for the many who may come to this text needing some revision the Appendix contains examples of algebraic and calculus methods which are relevant to reliability analysis.

Modelling of both component and system lifetime is taken to a level consistent with engineering or mathematical master's degrees, but it is possible to use the book as a more basic reference. Chapters 1 to 4 provide a grounding in the elements of modelling the lifetime of a single non-repairable unit. Reliability as a characteristic is a probability so Chapter 1 gives a guide to all the fundamentals of probability theory. The various measures commonly associated with reliability are also defined and described. No prior knowledge is assumed, but those for whom this is completely new may benefit from the support of some introductory probability and statistics text. In Chapter 2 the lifetime models in most common use are described and their essential characteristics discussed. In practice, the exponential, Weibull, normal, lognormal and gamma distributions cover a wide variety of situations.

Chapters 3 and 4 look at ways of choosing and fitting the most appropriate model to a given data set. There are two critical points to note here. First, the effect of censoring: censored data occur where a lifetime is only known to be less than or greater than a certain value, that is, failure has not been 'observed'. This can eliminate

some statistical measures. For example, average lifetime in the form
of the arithmetic mean cannot be calculated. Fortunately, much of
the most desirable methodology can be adjusted to take account of
censoring. Second, in reliability a prime interest is estimating lifetimes in the tail of the distribution, lifetimes which are exceeded,
say, on 95% or 99% of occasions. By definition, tail values are not
observed very often and this lack of data makes estimation of tail
values difficult. We need samples of hundreds rather than tens of
lifetimes in order to predict these points with great accuracy. Typically, there may be for a given data set little to choose between the
fit of Weibull, lognormal and gamma models, but considerable difference in tail values with a specified reliability.

One of the most difficult concepts to convey to students is the fundamental difference in the analysis of lifetimes for a repairable system
versus a non-repairable system. The phrase 'failure rate' is subject to
so much misconception. The key question for a repairable system is
whether repair truly 'renews' the system. Chapter 5 attempts to explain
and illustrate these key points. Chapter 6 introduces methods for dealing with systems where more than one component or subsystem and
their reliability characteristics are specified. The complexity of the analysis quickly increases with systems of many parts, so inevitably simpli-
fying assumptions start to appear. Chapter 7 demonstrates general
techniques for dealing with functions of more than one random variable
and includes the example of stress–strength modelling.

Many systems are repairable and issues relating to maintenance are
of great interest since maintenance contributes directly to reliability and
availability. It also tends to be a high-cost activity, so using it efficiently
is important. Chapter 8 discusses the effect of different types of maintenance strategy. Perhaps a key point, both here and in Chapter 5, is that
maintenance/repair does not improve reliability purely by definition.
Chapter 9 deals with more advanced approaches to the analysis of life
test data. This follows on from the estimation methods of Chapters 3
and 4 and will be mathematically challenging for some readers.

Finally, Chapter 10 draws together snapshot introductions to a
range of advanced models, often incorporating some extra variable
or parameter which reflects an additional dimension of the lifetime
population under consideration. Two case studies which illustrate
various ideas from throughout the book are given. For the examples
generally, the package MINITAB has been used as the principal
statistical tool for analysis and illustration.

Linda C. Wolstenholme
City University, London

CHAPTER 1

Basic Concepts

At the bottom of the theory of probabilities is only common sense expressed in numbers.

Essai philosophique des probabilités
– P. S. Laplace (1814)

1.1 Introduction

The term *reliability* usually refers to the probability that a component or system will operate satisfactorily either at any particular instant at which it is required or for a certain length of time. Fundamental to quantifying reliability is a knowledge of how to define, assess and combine probabilities. This may hinge on identifying the form of the variability which is inherent in most processes. If all components had a fixed known lifetime there would be no need to model reliability.

1.2 Events and probability

Most of us have some intuitive feel for the terms *probability* or *chance*, but not necessarily a very rational approach to their interpretation.

Let A be some *event* of interest – say, a software failure or material flaw. Let $P(A)$ be the probability that the event A occurs. The value of $P(A)$ may range from zero (the event is impossible) to one (the event is certain). An event is a particular *outcome*, or set of outcomes, of a particular trial or experiment. Triggering a switch could be a *trial*, and possible outcomes are 'works' or 'doesn't work'.

When events are distinct, in the sense that one event precludes all others, then the events are described as *exclusive*. This may be illustrated in terms of a Venn diagram. The universal set, Ω, depicted by a rectangle in the diagram, represents all possible outcomes, with A, B, C, etc. particular outcomes. If A and B are exclusive they will be

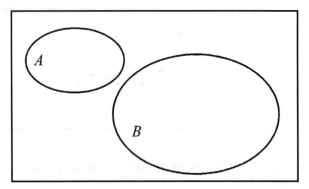

Figure 1.1. Venn diagram, with A and B exclusive

represented by non-overlapping sets (Figure 1.1). If there are more than two classes of outcome and they are all non-overlapping, then these events are *mutually exclusive*.

Classification of events and their relationship is important. For example, a component may have different ways in which it can fail. Knowledge of whether these modes are exclusive or whether there may be more than one cause of any individual failure affects the reliability. Similarly, it is important to know whether modes of failure are related.

Probabilities may be assigned to individual events in a variety of ways.

(i) We may know that the events are *equiprobable*, that is, each has the same chance of occurrence. If all possible outcomes, N in number, of a trial are equally likely to occur then each outcome has a probability of $1/N$.

(ii) *Past experience* or *experimentation* is often relied upon. If a large number of trials are carried out then an estimate of the true probability may be obtained. Suppose a large number, K, of an item are tested or inspected and m found to be satisfactory; then an estimate of the probability that an item is satisfactory is given by m/K. A potential problem with this is that K does need to be large, say greater than 50, and if testing is expensive, time-consuming or destructive, a good estimate of the probability may be difficult.

(iii) When a reliable estimate of a probability is not possible it may be a case of relying on past experience, intuition, or what is loosely called engineering judgement! Fortunately, there are ways in which this initial guesswork can be refined using subsequent experimentation, in particular via Bayesian techniques, which are introduced in Chapter 10.

The situation in (i) above is an example of using knowledge about the pattern of outcomes. These patterns may take many forms. The equiprobable pattern is a simple example. An appropriate model for characteristics like length of life or stress at failure is what is required, and specific models will be discussed in the next chapter.

1.3   Rules of probability

Regardless of how we assign values to the probabilities of individual events, the same rules apply to the combining of such probabilities, or the determination of related probabilities.

A fundamental principle is that when mutually exclusive events are put together, the probability that the outcome of a trial belongs to this set is the sum of the probabilities of the individual events, that is,

$$\text{for } \{A_i\} \text{ mutually exclusive, } P(A_1 \text{ or } A_2 \text{ or } \dots \text{ or } A_n) = \sum_{i=1}^{n} P(A_i).$$

The basic *axioms of probability* may be summarized as follows:

I. $0 < P(A) \leq 1$ for any event A.

II. $P(\Omega) = 1$.

III.   For $\{A_i\}$ mutually exclusive, $P(\cup A_i) = \sum P(A_i)$. ($\cup$ is the symbol *union* and $\cup A_i$ means that any one or more of the events occur.)

The event $\bar{A}$ is all outcomes which are not A, and therefore with A covers all possible outcomes. So,

$$P(A \text{ or } \bar{A}) = P(\Omega) = 1.$$

$\bar{A}$ is called the *complement* of A. Since A and $\bar{A}$ are exclusive,

$$P(A \text{ or } \bar{A}) = P(A) + P(\bar{A}),$$

so that

$$P(\bar{A}) = 1 - P(A). \tag{1.1}$$

Now consider two events A and B, which are not in general exclusive. However, the event A can be considered to have two exclusive parts, outcomes which are in A but not in B (event $A \cap \bar{B}$, where $\cap$ is the symbol *intersection*), and outcomes which are in both A and B (event $A \cap B$) (Figure 1.2):

$$A = (A \cap \bar{B}) \cup (A \cap B),$$

and therefore

$$P(A) = P(A \cap \bar{B}) + P(A \cap B). \tag{1.2}$$

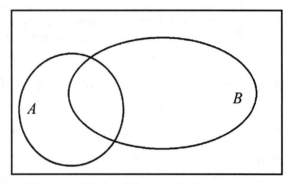

Figure 1.2. Venn diagram, with A and B not exclusive

When A and B are not exclusive the event $A \cup B$ has three exclusive parts, $A \cap \bar{B}$, $A \cap B$, and $\bar{A} \cap B$. So

$$P(A \cup B) = P(A \cap \bar{B}) + P(A \cap B) + P(\bar{A} \cap B).$$

Using Equation (1.2),

$$P(A \cap \bar{B}) = P(A) - P(A \cap B),$$

$$P(\bar{A} \cap B) = P(B) - P(A \cap B).$$

Therefore

$$P(A \cup B) = P(A) + P(B) - P(A \cap B). \tag{1.3}$$

When A and B are non-overlapping $P(A \cap B) = 0$, and (1.3) follows directly from Axiom III.

1.4   Dependent events

The probability of an event may be affected by the occurrence or non-occurrence of some other event. In many systems the failure of some components will result in extra load on others, thus accelerating their tendency to susequently fail.

If the probability of event A is a function of whether or not event B has occurred, then A is *dependent* on B. The probability of A under the condition that B occurs is denoted $P(A|B)$, often described as the probability of A *given* B. This is termed a *conditional probability*.

The probability of events A and B both occurring is $P(A \cap B)$. This involves a product of probabilities, as illustrated by the following example. Suppose for a given weapon system 99% of firings are successful and of those successfully fired 1 out of 11 directly hit the target. On average, out of 100 firings, 99 will launch and 9 will make a direct hit so the probability of a direct hit (which necessarily implies a successful launch) is 9/100.

To put this mathematically, let event A be 'firing successful' and event B 'direct hit'. Then

$$P(A \cap B) = P(A)P(B|A)$$
$$= \frac{99}{100}\frac{1}{11} = \frac{9}{100}.$$

When A and B are *statistically independent*, that is, the occurrence of one event does not affect the occurrence of the other,

$$P(B|A) = P(B) \text{ and } P(A \cap B) = P(A)P(B).$$

It is worth pointing out that the properties of exclusiveness and independence are quite different. They can sometimes be confused. It can be noted that while exclusivity may be demonstrated via a Venn diagram, it is not possible to illlustrate independence in this way.

So our final basic rule is

$$P(A \cap B) = P(A)P(B|A) = P(B)P(A|B). \qquad (1.4)$$

Rules (1.2) and (1.4) may be combined to express $P(A)$ in terms of conditional probabilities. From (1.4),

$$P(A \cap B) = P(B)P(A|B)$$

and

$$P(A \cap \bar{B}) = P(\bar{B})P(A|\bar{B}).$$

Substituting into (1.2),

$$P(A) = P(B)P(A|B) + P(\bar{B})P(A|\bar{B}).$$

This may be generalized to what will be termed the *theorem of total probability*. Let B_1, B_2, ..., B_n be a set of mutually exclusive events which completely overlap an event A. Then

$$P(A) = \sum_{i=1}^{n} P(B_i)P(A|B_i). \qquad (1.5)$$

Example 1.1: A component part for an electronic item is manufactured at one of three different factories, and then delivered to the main assembly line. Of the total number supplied, factory A supplies 50%, factory B 30%, and factory C 20%. Of the components manufactured at factory A, 1% are faulty, and the corresponding proportions for factories B and C are 4% and 2%, respectively. A component is picked at random from the assembly line. What is the probability that it is faulty?

Let Y denote the event 'supplied by factory Y', and F denote the event 'part is faulty'. Then the following probabilities are known:

$$P(A) = 0.5, \quad P(B) = 0.3, \quad P(C) = 0.2;$$
$$P(F|A) = 0.01, \ P(F|B) = 0.04, \ P(F|C) = 0.02.$$

We require $P(F)$, which is the *unconditional* probability that a component is faulty. By the theorem of total probability,

$$P(F) = \sum_{i=1}^{3} P(F|Y_i)P(Y_i),$$

that is,

$$P(F) = P(F|A)P(A) + P(F|B)P(B) + P(F|C)P(C)$$
$$= 0.01 \times 0.5 + 0.04 \times 0.3 + 0.02 \times 0.2$$
$$= 0.021.$$

An important theorem in the calculation of conditional probabilities is Bayes' theorem. From (1.4),

$$P(A)P(B|A) = P(B)P(A|B).$$

So

$$P(B|A) = \frac{P(B)P(A|B)}{P(A)}.$$

Taking B to be any event B_j from a set of mutually exclusive events completely covering A and invoking (1.5), we have

$$P(B_j|A) = \frac{P(B_j)P(A|B_j)}{\sum_{i=1}^{n} P(B_i)P(A|B_i)}, \tag{1.6}$$

which is the formula attributed to Rev. Thomas Bayes (1702–1761).

Example 1.2: In Example 1.1, what is the probability that a component found to be faulty came from factory A?

We require

$$P(A|F) = \frac{P(A)P(F|A)}{P(A)P(F|A) + P(B)P(F|B) + P(C)P(F|C)}$$
$$= \frac{0.005}{0.005 + 0.012 + 0.004}$$
$$= 0.24.$$

1.5   Random variables and probability distributions

A *random variable* X may be used to numerically represent the outcome of an experiment. A random variable is called *discrete* if it can assume only a finite number, or an infinite sequence, of distinct values. Frequently this means that X takes only (positive) integer values. Experiments involving counting or scoring are typical sources of discrete random variables.

The use of random variables enables us to put questions of probability into a mathematical framework and hence to apply mathematical techniques. In this text upper-case letters will be used for the *name* of a random variable and lower-case letters for a particular *value* of the random variable. As stated in Section 1.2, it will be of interest to know something about the pattern of outcomes, that is, the way in which total probability (value 1) is distributed over all possible outcomes.

Let X be a discrete random variable. A *probability distribution* will be defined by the function

$$p_X(x) = P(X = x).$$

This may be presented in the form of a set of separate numeric values, or, in the case of well-defined patterns of probability distribution, as an algebraic function. All forms must satisfy $\sum_x p(x) = 1$.

> **Example 1.3:** Take the production line problem of Example 1.1 again. For all components, $P(A) = 0.5$, $P(B) = 0.3$ and $P(C) = 0.2$. But suppose we now want the distribution of factory source for non-faulty items only. This distribution is conditional on $\bar{F}$. Using Bayes' theorem in the way shown in Example 1.2, we find $P(A|\bar{F}) = 0.506$, $P(B|\bar{F}) = 0.294$ and $P(C|\bar{F}) = 0.200$.

> Let the random variable X take the values 1, 2, 3 to represent the factory source of a faulty item. Then

x:	1	2	3
$p_X(x)$:	0.506	0.294	0.200

As is expected, this probability distribution sums to 1. The subscript X on $p(x)$ will now be dropped but will be used wherever the intention is less clear.

Example 1.4: A number of 'one-shot' devices are to be tested and the probability that the device will fail is defined to be θ. Of interest is the distribution of the number of tests until the first failure.

Let X be a random variable representing the number of tests up to and including the first failure. The probability that X takes the value x is given by

$p(x) = P$(first x–1 tests successful and xth test fails)

$\quad = P$(1st test successful)P(2nd test successful) ... P(xth test fails)

by the product rule of probabilities (1.4), and since all tests are independent,

$$p(x) = (1 - \theta)(1 - \theta)...(1 - \theta)\theta$$
$$= (1 - \theta)^{x-1}\theta.$$

The distribution of probability will vary with the value taken by θ, and θ is called a *parameter* of the distribution. The algebraic form of $p(x)$ defines a *family of distributions* and the value of θ identifies a particular member of the family. There may be more than one parameter, so it is possible to generalize and consider a probability distribution to be conditional on a vector of parameters, $\boldsymbol{\theta}$.

Figure 1.3 illustrates a discrete probability distribution in the form of a *histogram* of theoretical probabilities, where the heights of the bars are proportional to $p(x)$ for each discrete value of the variable x.

A random variable is termed *continuous* when it may assume any value in some interval. Variables representing measurements such as length, time, temperature, stress etc. are considered to be continuous, even though in practice the rounding of measurements means that only a subset of values in the range may be recorded.

For a continuous random variable, T, the probability distribution is defined via a function called the *probability density function* which outlines the distribution of probability over the range of the random variable. This function, which will be called here $f(t)$, is necessarily defined theoretically in algebraic terms. It does not yield probabilities directly in the way that $p(x)$ does for discrete random variables, but does so via integration.

Many continuous random variables, especially those appropriate in reliability, take only positive values. A typical $f(t)$ may be similar

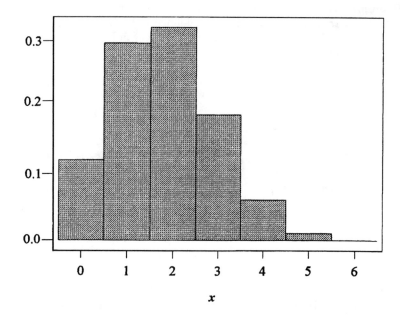

Figure 1.3. Histogram of theoretical probabilities for a discrete probability distribution

to that shown in Figure 1.4. The distribution of the area between $f(t)$ and the t axis characterizes the probability distribution, namely,

$$P(T \leq t) = \int_0^t f(u)\mathrm{d}u \, .$$

More generally, we can write

$$P(t_1 < T \leq t_2) = \int_{t_1}^{t_2} f(t)\mathrm{d}t \, .$$

By definition, $f(t)$ must be everywhere positive over the range of T, and the total area under $f(t)$ over the range of T must be 1, since this represents total probability.

Example 1.5 : The lifetime of a certain component is considered to have a distribution of the form

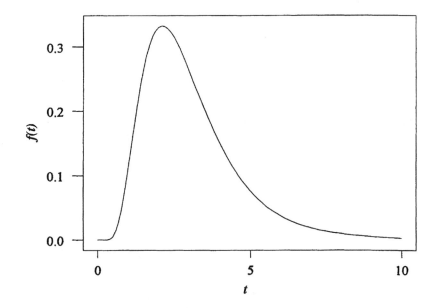

Figure 1.4. A probability density function

$$f(t) = 2te^{-t^2}, \quad t \geq 0.$$

The probability that the lifetime lies in the range t_1 to t_2 is given by

$$\int_{t_1}^{t_2} 2te^{-t^2} dt = e^{-t_1^2} - e^{-t_2^2}.$$

Putting $t_1 = 0$ and $t_2 = \infty$ gives a total probability of one, verifying that the form of $f(t)$ is a proper probability density function.

1.6   The reliability function

Let T be a continuous random variable representing a lifetime characteristic, say time to failure, with probability density function $f(t)$. It is assumed that T is non-negative, and that an origin and scale of measurement are defined. A particular realization of T is denoted t.

The *distribution function* is given by

$$F(t) \;=\; P(T \le t) \;=\; \int_{0}^{t} f(u)\mathrm{d}u \,.$$

$F(t)$ describes the accumulation of failure probability as t increases. By definition, $F(t)$ is increasing, is zero at $t = 0$ and tends to one as t tends to infinity [Figure 1.5]. It also follows that $f(t)$ can be obtained from $F(t)$ by differentiation:

$$f(t) \;=\; \frac{\mathrm{d}}{\mathrm{d}t}[F(t)] \,, \text{ denoted } F'(t).$$

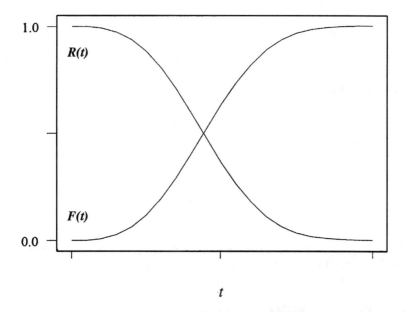

Figure 1.5. The distribution and reliability functions

The *p*th *quantile*, or 100*p*th *percentile* of the distribution of T is the value t_p such that

$$P(T \le t_p) = F(t) = p.$$

Such points in a lifetime distribution are relevant to, for example, the guaranteed lifetime of a consumer product.

The *reliability function* or *survivor function*, $R(t)$, is given by

$$R(t) = 1 - F(t) = P(T > t).$$

This is the probability that the lifetime exceeds t and is the principal reliability measure. We say that

$$R(t_0) = P(T > t_0) = \int_{t_0}^{\infty} f(t)\mathrm{d}t$$

is the *reliability at* t_0. The reliability function is complementary to $F(t)$, taking the value one at $t = 0$ and tending to zero as t tends to infinity [Figure 1.5]. $F(t)$ and $R(t)$ coincide when both functions equal 0.5. The value of t at this point, $t_{0.5}$, is the *median*, which is one possible measure of average lifetime.

Example 1.6: A product which has reliability function

$$R(t) = \frac{2}{2 + t^3},$$

where t is measured in years, carries a six-month guarantee. The probability that the product fails in the guarantee period is given by

$$1 - R(0.5) = 1 - \frac{2}{2 + (1/2)^3} = 0.0588.$$

To determine the length of guarantee period necessary to give a failure probability of say, 0.01, the value of $t_{0.01}$ is required and given by solving

$$0.99 = \frac{2}{2 + t_{0.01}^3}.$$

So,

$$t_{0.01} = \left(\frac{2}{0.99} - 2 \right)^{1/3} = 0.272 \text{ years}.$$

Therefore an appropriate guarantee period for this product might be three months only.

In reliability analysis the *mean time to failure* (MTTF) is often of interest. This is given by

$$\mu = \int_0^\infty tf(t)dt. \tag{1.7}$$

Now we can show that when T is defined on $[0, \infty]$ the MTTF, μ, is the area between $R(t)$ and the t axis. This provides a useful comparison of different reliability functions. Evaluating the right-hand side of (1.7) by means of intergration by parts, we obtain

$$\mu = [tF(t)]_0^\infty - \int_0^\infty F(t)dt$$

$$= [t(1 - R(t))]_0^\infty - \int_0^\infty (1 - R(t))dt$$

$$= [t(1 - R(t)) - t]_0^\infty + \int_0^\infty R(t)dt$$

$$= [tR(t)]_0^\infty + \int_0^\infty R(t)dt.$$

In the product $tR(t)$, $R(t)$ tends to zero as t tends to infinity, more quickly than t tends to infinity, therefore the first term of μ is zero, yielding

$$\mu = \int_0^\infty R(t)dt. \tag{1.8}$$

In Figure 1.6 the area under $R_2(t)$ is clearly larger than that under $R_1(t)$ and is accompanied by higher reliability over all t. In Figure 1.7 the lifetime distributions have the same MTTF but are very different in nature. The choice as to which was the better model in a given application would depend on a number of things. A principal factor

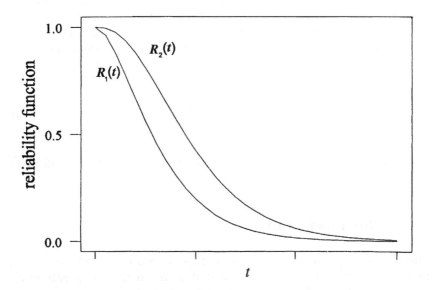

Figure 1.6. R_2 has a larger MTTF than R_1

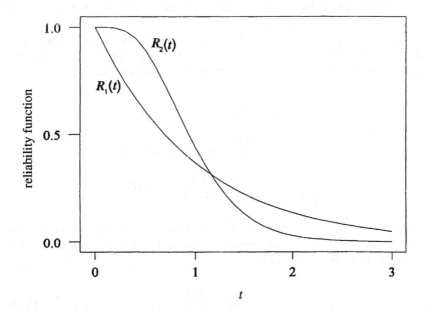

Figure 1.7. Two reliability functions with the same MTTF (= 1)

would be the required life of the product. Clearly for low values of t, $R_2(t)$ is more satisfactory. However, with this model, once reliability begins to fall it does so very rapidly. The distinguishing factor between these models is not the MTTF but the *variance*. The variance measures the degree to which the lifetime distribution is spread out. It is given by

$$\sigma^2 = \int_0^\infty (t - \mu)^2 f(t) \mathrm{d}t \tag{1.9}$$

The *standard deviation* is σ, the square root of the variance, and takes the same units as t.

1.7   The hazard function

The probability density function is proportional to the unconditional probability of failure at time t, but it is more useful in reliability analysis to look at how prone an item is to failure, having survived to time t.

Consider a small interval of time $[t, t+\delta t]$. The unconditional probability that a unit fails in this interval is $R(t) - R(t+\delta t)$. For very small δt, this is approximately $f(t)\delta t$.

Let event A be 'surviving beyond t' and let event B be 'failing in $[t, t+\delta t]$'. Clearly event A includes event B. The probability that the unit fails in $[t, t+\delta t]$, given that it has not failed in $[0, t]$ is

$$P(B|A) = \frac{P(A \cap B)}{P(A)}, \quad \text{from (1.4),}$$

$$= \frac{P(B)}{P(A)}, \quad \text{since } A \text{ includes } B,$$

$$\approx \frac{f(t)\delta t}{R(t)} = h(t)\delta t.$$

The function $h(t)$ is called the *hazard function*, or *instantaneous/age-specific failure rate*. It describes how prone the unit is to failure after a length of time. Note that

$$h(t) = \frac{f(t)}{R(t)}. \tag{1.10}$$

The *cumulative hazard function* is

$$H(t) = \int_0^t h(u)\,du$$

$$= \int_0^t \frac{f(u)}{R(u)}\,du = \int_0^t \frac{-R'(u)}{R(u)}\,du$$

$$= [-\log R(u)]_0^t = -\log R(t).$$

Therefore

$$R(t) = \exp[-H(t)].$$

It is only necessary to know one of the functions, $h(t)$, $f(t)$, $R(t)$ in order to be able to deduce the other two, as demonstrated in Figure 1.8.

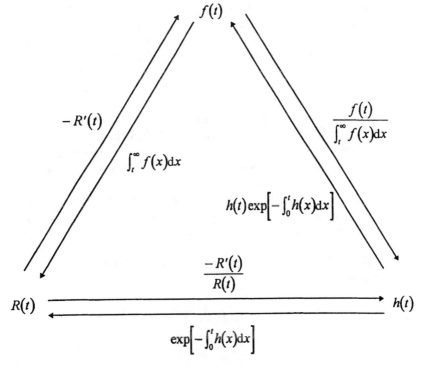

Figure 1.8. The relationship between $h(t)$, $f(t)$ and $R(t)$

The hazard function is important because it has a direct physical interpretation, and information about the nature of the function is useful in selecting an appropriate model for lifetime.

The hazard function may take a variety of forms:

(i)   Constant: $h(t) = \lambda$. Therefore $H(t) = \lambda t$, and $R(t) = e^{-\lambda t}$. This is the reliability function of the *exponential distribution* with rate parameter λ. An exponential life distribution corresponds to 'no ageing'. It is as though a unit is 'as new' at each instant of time. This is known as the *memoryless* property. For example, an electronic device may be subjected to some environment which is a random process, such as a power surge or other 'shock'. If the device fails when a shock occurs, but not otherwise, then the time between shocks represents the time to failure of the device.

(ii) $h(t)$ is an increasing function of t. The unit is subject to ageing through wear, fatigue, or accumulated damage. In practice, this is the most common case.

(iii) $h(t)$ is a decreasing function of t. This is less common, but may be true, in part, of a manufacturing process where low-quality components are likely to fail early. A 'burn-in' process may be used to remove these defective items, leaving the higher-quality components which then exhibit gradual ageing. Similarly, a mechanical device may require 'case-hardening', a period of being 'worked' or 'run in', in order to harden the device, after which it becomes more reliable. The complete picture is given by the so-called:

(iv) 'Bathtub' hazard. Here we have an initial decreasing hazard, followed by a fairly constant period, called the 'useful life', and a final phase, 'wear-out', where the hazard rate increases [Figure 1.9]. It is not usually beneficial to model the complete bathtub curve in a sophisticated way. Often the different phases may be treated separately.

Example 1.7: A piecewise linear model may be an adequate approximation for a bathtub hazard. Let $h(t)$ be given by

$$h(t) = \begin{cases} \lambda + c_0(t_0 - t), & 0 \leq t \leq t_0 \\ \lambda, & t_0 \leq t \leq t_1 \\ \lambda + c_1(t - t_1), & t > t_1 \end{cases}$$

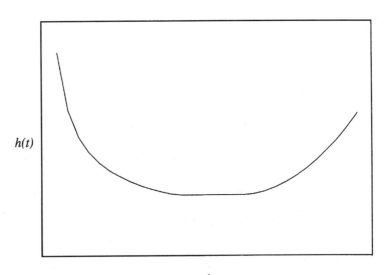

$h(t)$

t

Figure 1.9. A bathtub hazard function

with c_0 and c_1 positive. This function models a hazard taking the value $\lambda + c_0 \, t_0$ at $t = 0$, decreasing linearly until $t = t_0$, staying constant until $t = t_1$ and increasing linearly thereafter.

1.8   Expectation

The measures described in (1.7) and (1.9) have an important bearing on the location, shape and spread of a probability distribution. They are examples of distribution *moments* obtained via the general formulation of *expected values*.

Let some function of a continuous random variable X be given by $u(X)$. Then the expected value of $u(X)$ is given by

$$E[u(X)] = \int_X u(x)f(x)\mathrm{d}x. \tag{1.11}$$

There is an analogous definition in the case of discrete X, namely

$$E[u(X)] = \sum_X u(x)p(x). \tag{1.12}$$

The distribution mean and variance, (1.7) and (1.9), can be expressed in terms of expected values: the mean μ is

$$\mu = E(X) \qquad (1.13)$$

and σ^2, denoted $V(X)$, is given by

$$V(X) = E(X^2) - \mu^2 = E(X^2) - [E(X)]^2. \qquad (1.14)$$

Expected values may involve some condition, such as the factory source giving rise to the probability distribution in Example 1.3. The *mean residual lifetime* measures the mean remaining lifetime for an item that has survived so far. If the current age is t_0 and the age at failure is T then the mean residual lifetime is given by

$$m(t_0) = E(T - t_0 \mid T \geq t_0).$$

It can be shown that this conditional expected lifetime is given by

$$m(t_0) = \frac{\int_{t_0}^{\infty} R(x)\,dx}{R(t_0)}.$$

By definition $m(0) = \mu$, and if a distribution has $m(t_0) \leq m(0)$ for all $t_0 \geq 0$, it is said to have the 'new better than used' property.

CHAPTER 2

Common Lifetime Models

2.1 Introduction

Reliability is the science of predicting, estimating, or optimizing the life distribution of *components* or *systems*. The reliability of a product is a vital part of quality as perceived by the customer. It is therefore important for manufacturers to have precise information about the reliability of their products, both at the production stage and subsequently from performance in the field. Issues of safety, product liability and warranties are all strongly dependent on reliability. Statistical theory is used in the design and analysis of life tests and in the collection and analysis of data from the field.

Systems may fail for a variety of reasons. Such causes as design faults, operating error and poor maintenance are essentially under human control, but other failures may be more or less random and unpredictable. Many systems may be broken down into components and each component may have several possible states, other than just failed or working.

The reliability of a system is characterized by the individual component reliabilities and by the contributions which individual components make to the reliability of the system. To improve system reliability, individual component reliabilities may be improved, or the system redesigned to make the contributions of individual components less critical. At this point in considering models for lifetimes a distinction between systems and components will not necessarily be made, the subject of the discussion will simply be referred to as a 'unit'. Reliability is measured either in terms of probability of failure or in terms of the loss, that is, the consequence of failure. The loss may be purely financial, but other factors may be included, such as human safety.

Reliability data are usually measurements of some variable to do with failure, such as time to failure, or load at failure, and possibly additional measurements on a group of units. We are concerned with the problem of estimating reliabilities and the prediction of reliability on the basis of these observations. The aim of statistical reliability analysis is to turn reliability data into information about the population of units that is the basis for action to change that population (Figure 2.1).

2.2   The Poisson process

Failures which occur in some continuous medium, such as time, are said to occur as a *Poisson process* if the probability of a failure occurring at any time is constant and failures are independent. This is the notion of failures occurring 'at random'. The constant hazard rate gives rise to the exponential distribution [Section 1.7] which has probability density function

$$f(t) = \lambda e^{-\lambda t}, \ t \geq 0$$

and reliability function

$$R(t) = e^{-\lambda t}.$$

If it is the *number of failures* in a given time interval which is of interest then the probability distribution is defined by

$$P(X = x) = \frac{e^{-\mu}\mu^x}{x!}, x = 0, 1, 2, \dots ,$$

where X is the discrete random variable representing the number of failures in time t, and μ is the expected number of failures in time t, that is, $\mu = \lambda t$. This is called the *Poisson distribution*.

To demonstrate the consistency of this model with the exponential model for the time between events, it is noted that

$$R(t) = P(\text{no failure in time } t)$$

$$= P(X = 0) = \frac{e^{-\mu}\mu^0}{0!} = e^{-\lambda t}.$$

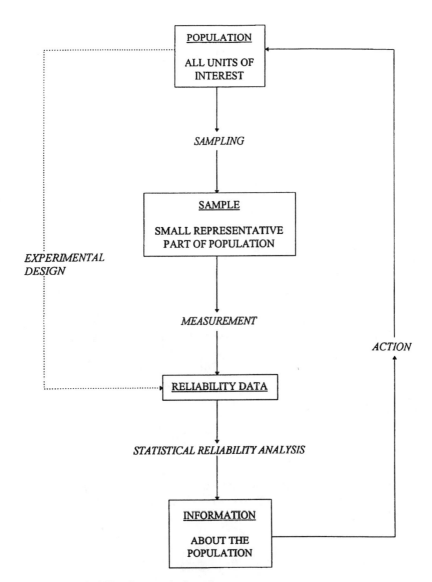

Figure 2.1. Reliability data analysis cycle

In some sense the exponential model is fairly restrictive with its constant hazard rate assumption, but it does turn out to be a representative model in many situations. In some of the distribution families now to be considered the exponential model proves to be a special case.

2.3   The Weibull distribution

This is one of the most widely used lifetime distributions. It is justified
on a number of grounds, but perhaps its prime feature is that it has
a versatile power law for its hazard function. It has *scale parameter*
α and *shape parameter* β. The scale parameter has the same units as
the random variable but the shape parameter has no units – it is
dimensionless. The Weibull distribution is defined by

$$R(t) = \exp[-(t/\alpha)^\beta],\ t \geq 0$$

$$f(t) = \frac{\beta}{\alpha}\left(\frac{t}{\alpha}\right)^{\beta-1} \exp[-(t/\alpha)^\beta],$$

and hence

$$h(t) = c\ t^{\beta-1},$$

where $c = \beta/\alpha^\beta$. The distribution has an increasing hazard rate if $\beta >$
1 and a decreasing hazard rate if $\beta < 1$; $\beta = 1$ gives the constant hazard
of the exponential distribution (Figure 2.2).

To find the mean and variance of the Weibull distribution we need
the $E(T^r)$, $r = 1, 2$, given by $\alpha^r\ \Gamma(1 + r/\beta)$, where

$$\Gamma(a) = \int_0^\infty x^{a-1}e^{-x}dx$$

is the so-called *gamma function*. A useful special case of $\Gamma(a)$ arises when
a is a positive integer. Then $\Gamma(a) = (a-1)!$. In all cases $\Gamma(a+1) = a\Gamma(a)$.

It then follows from (1.13) that the mean, μ, is given by

$$\mu = \alpha\Gamma\left(1 + \frac{1}{\beta}\right).$$

For very small and very large values of β the gamma function is
approximately 1. For other values of β the function lies between 1 and
approximately 0.8. Due to the unfriendly nature of the gamma function
the value of α is often thought of as a measure of average in preference
to the mean. It is frequently referred to as the *characteristic lifetime*.

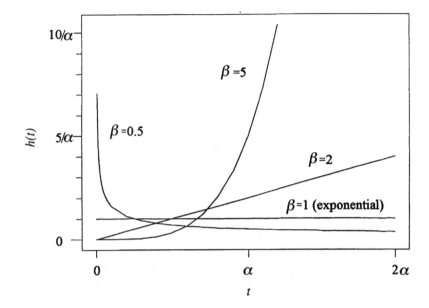

Figure 2.2. Weibull hazard functions

$R(\alpha)$ is $\exp[-1] = 0.368$ and therefore α is approximately the 0.63 quantile of the distribution.

From (1.14) the variance is

$$E(T^2) - [E(T)]^2 = \alpha^2 \Gamma\left(1 + \frac{2}{\beta}\right) - \alpha^2 \Gamma^2\left(1 + \frac{1}{\beta}\right).$$

Once again this expression is not user-friendly, but general statements may be made regarding its nature. First, it is proportional to α^2. Second, it can be shown that the variance is inversely related to the value of β. That is, the larger the value of β, the lower the variability in the lifetime (Figure 2.3).

The Weibull distribution also arises as an asymptotic *extreme value* distribution (see Section 2.9). Suppose that failure time is the smallest of a large number of independent and identically distributed (IID) non-negative random variables; then it can be shown that, under certain conditions, the limiting distribution of failure time is Weibull distributed. A special exact case is summed up in the important *weakest-link* property, an example of which is given below.

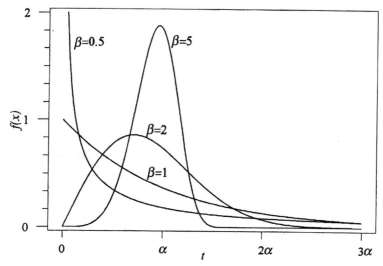

Figure 2.3. Weibull probability density functions

Suppose a system has a series or 'chain' of n components, each of which has failure time with Weibull distribution, parameters α and β. Let Y be the failure time of the system. Then

$$R_Y(y) = P(Y > y) = P(\text{all components have } T > y) = [R_T(y)]^n,$$

using the probability product rule (1.4) for independent events. So,

$$R_Y(y) = [\exp[\,-(y/\alpha)^\beta\,]]^n = \exp[\,-n(y/\alpha)^\beta\,]$$

$$= \exp[\,-\{y/(\alpha n^{-1/\beta})\}^\beta].$$

Thus it can be seen that Y also has a Weibull distribution, with the same shape parameter, β, as T, but the scale parameter is $\alpha n^{-1/\beta}$. This demonstrates that, provided $\beta > 1$, a long chain of components has a lower time to failure than a short chain.

Example 2.1: The life (in years) of certain generator field windings is approximated by a Weibull distribution with $\alpha = 13$ years and $\beta = 2$. The reliability for a two-year warranty is given by

$$R(2) = \exp[-(2/13)^2] = 0.9766.$$

Example 2.2: Carbon fibres of length 100 mm are estimated to have Weibull distributed strength with $\alpha = 3.85$ GPa and $\beta = 31$. If the weakest-link property applies the strength of fibres of differing lengths will also be Weibull distributed with the same shape parameter $\beta = 31$ but scale parameter given by

$$\alpha_1 = \alpha\left(\frac{l}{100}\right)^{-1/\beta} = \alpha\left(\frac{100}{l}\right)^{1/\beta}.$$

Fibres of length 4 mm will have characteristic strength, that is, scale parameter $\alpha_4 = 3.85(25)^{1/31} = 4.27$. Fibres of length 1 mm will have scale parameter $\alpha_1 = 3.85(100)^{1/31} = 4.47$. Thus the shorter the fibres, the stronger they are.

2.4   The Gumbel distribution

The Gumbel distribution has reliability function

$$R(t) = \exp\left[-\exp\frac{(t-\gamma)}{\eta}\right], \qquad -\infty < t < \infty,$$

where γ is the *location parameter* and η the scale parameter. This distribution is used to model lifetimes even though its range includes negative values. If the value of γ is suitably large, the probability of negative lifetimes is negligible.

The hazard function has an exponential form:

$$h(t) = \frac{f(t)}{R(t)} = \frac{-R'(t)}{R(t)} = \frac{\frac{1}{\eta}\exp\left(\frac{t-\gamma}{\eta}\right)\exp\left[-\exp\left(\frac{t-\gamma}{\eta}\right)\right]}{\exp\left[-\exp\left(\frac{t-\gamma}{\eta}\right)\right]}$$

$$= \frac{1}{\eta}\exp\left(\frac{t-\gamma}{\eta}\right).$$

One particularly useful feature of the Gumbel distribution is that it models log lifetimes when the lifetimes are Weibull distributed. Let

$Y = \log T$, where T has a Weibull distribution with parameters α and β. Then

$$
\begin{aligned}
R_Y(y) &= P(Y > y) = P(\log T > y) = P(T > e^y) \\
&= R_T(e^y) \\
&= \exp\left[-\left(\frac{e^y}{\alpha}\right)^{\beta}\right] = \exp[-(e^y/e^\gamma)^{1/\eta}] \\
&= \exp[-\exp([y - \gamma]/\eta)].
\end{aligned}
$$

The scale parameter of the Gumbel distribution is a function of the Weibull shape parameter and the location parameter is a function of the Weibull scale parameter:

$$\eta = 1/\beta, \; \gamma = \log \alpha.$$

The Gumbel distribution has no shape parameter because it has a fixed shape (Figure 2.4).

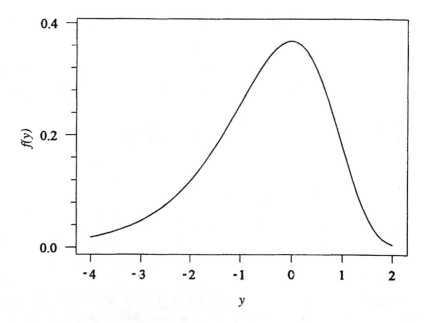

Figure 2.4. The standardized Gumbel distribution ($\gamma = 0$, $\eta = 1$)

2.5 The normal and lognormal distributions

The familiar bell-shaped model with probability density function

$$f(t) = \frac{1}{\sigma\sqrt{2\pi}}\exp\left[-\frac{1}{2}\left(\frac{t-\mu}{\sigma}\right)^2\right], \quad -\infty < t < \infty,$$

was first defined by the eighteenth-century mathematician Gauss, but is commonly referred to as the normal distribution. It forms the backbone of much statistical analysis and, in particular, has a special role as a limiting distribution. It is not, however, widely appropriate as a lifetime distribution. The fact that negative t values are allowed is not a special problem. As with the Gumbel distribution, if μ and σ are set appropriately the probability of a negative t becomes negligibly small. The practicality is that lifetime distributions tend to be asymmetric, with a long right-hand tail. An often occurring case is that of the lognormal distribution which models a variate whose logarithm is normally distributed.

Let $Y = \log T$ be a normally distributed random variable with mean μ and variance σ^2. The distribution function for Y is

$$P(Y < y) = F_Y(y) = \int_{-\infty}^{y} f_Y(x)\mathrm{d}x.$$

This integral cannot be evaluated in closed form but is obtained for particular y from normal probability tables. Values of $F_Y(y)$ are given by $\Phi(z)$ from the tables, with $z = (y - \mu)/\sigma$.

Now $T = \exp(Y)$ has a distribution defined by

$$P(T < t) = F_T(t) = P(\mathrm{e}^Y < t) = P(Y < \log t)$$

$$= \int_{-\infty}^{\log t} \frac{1}{\sigma\sqrt{2\pi}} \exp\left[-\frac{1}{2}\left(\frac{x-\mu}{\sigma}\right)^2\right]\mathrm{d}x.$$

Substituting $u = \mathrm{e}^x$ yields

$$F_T(t) = \int_{-\infty}^{t} \frac{1}{\sigma u\sqrt{2\pi}} \exp\left[-\frac{1}{2}\left(\frac{\log u - \mu}{\sigma}\right)^2\right]\mathrm{d}u,$$

which in turn demonstrates that the probability density function of t is

$$f_T(t) = \frac{1}{\sigma t \sqrt{2\pi}} \exp\left[-\frac{1}{2}\left(\frac{\log t - \mu}{\sigma}\right)^2\right], \quad 0 < t < \infty.$$

This describes the lognormal distribution. It is an exponential transformation of the normal distribution. If lifetimes have a lognormal
distribution then the log lifetimes have a normal distribution. The
lognormal distribution is to the normal distribution as the Weibull
distribution is to Gumbel. Note that the parameters μ and σ^2 of the
lognormal are the mean and variance of the log lifetime distribution.
Like the Gumbel, the normal distribution has location and scale
parameters only, and like the Weibull, the lognormal is a distribution
of varying shape (Figure 2.5).

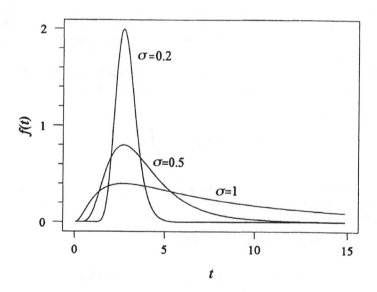

Figure 2.5. Lognormal probability density functions ($\mu = 1$)

The lognormal distribution is difficult to deal with analytically, and
also has a particular disadvantage in the form of its hazard function.
Initially, $h(t)$ increases, reaches a maximum and then slowly decreases,
tending to zero as $t \to \infty$. However, when large values of t are not of
interest, the model is often found to be suitable. The distribution can also
be derived theoretically for such processes as the growth of fatigue cracks.

2.6   The gamma distribution

The gamma distribution has a particular application as the model
arising for the sum of IID exponential variables, and can be derived
for some other failure processes. The practical usefulness of the distribution is inhibited by the fact that its reliability and hazard functions cannot in general be expressed in closed form.

The probability density function is given by

$$f(t) = \frac{\lambda(\lambda t)^{\alpha-1}e^{-\lambda t}}{\Gamma(\alpha)}, \; t \geq 0.$$

The shape parameter, α, controls the behaviour of the hazard
function: $0 < \alpha < 1$ gives a decreasing hazard rate and $\alpha > 1$ an
increasing rate. The case $\alpha = 1$ is the exponential distribution. The
parameter λ is the scale parameter and, in the case where the gamma
distribution is modelling the behaviour of a sum of IID exponential
random variates, represents the exponential hazard rate. In this case
α is the number of variates in the sum (Figure 2.6).

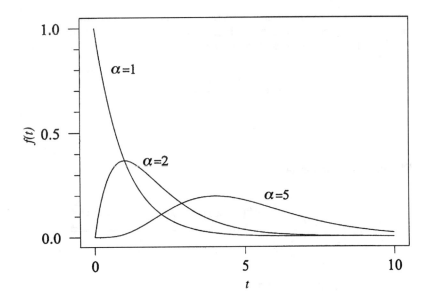

Figure 2.6. Gamma probability density functions ($\lambda = 1$)

Example 2.3: Suppose at $t = 0$ a unit is set in operation, and when it fails it is replaced by a spare, and so on until $n - 1$ spares have been used. If all units have exponentially distributed lifetimes, rate λ, then the time to the last failure has a gamma distribution with parameters λ and $\alpha = n$.

The reliability function $R(t)$ represents the probability that the time to the nth event is at least t. This is equivalent to the probability that there are at most $n - 1$ events in time $[0, t]$. The form of $R(t)$ may be deduced from Poisson probabilities as follows. We have that

$$P(i \text{ events in } [0, t]) = e^{-\lambda t}\frac{(\lambda t)^i}{i!}.$$

Then

$$R(t) = P(\text{ no more than } n - 1 \text{ events in } [0, t])$$

$$= e^{-\lambda t}\sum_{i=0}^{n-1}\frac{(\lambda t)^i}{i!}.$$

Example 2.4: Example 2.3 describes a method of system protection called *standby redundancy*. Consider a three-unit standby system where all units have failure rate λ. From above

$$R(t) = e^{-\lambda t}\sum_{i=0}^{2}\frac{(\lambda t)^i}{i!}$$

$$= e^{-\lambda t}\left[1 + \lambda t + \frac{(\lambda t)^2}{2}\right]$$

The hazard function is

$$h(t) = \frac{-R'(t)}{R(t)} = -\frac{e^{-\lambda t}[\lambda + \lambda^2 t] + [-\lambda e^{-\lambda t}]\left[1 + \lambda t + \frac{(\lambda t)^2}{2}\right]}{e^{-\lambda t}\left[1 + \lambda t + \frac{(\lambda t)^2}{2}\right]}$$

$$= \frac{\lambda^3 t^2}{2 + 2\lambda t + \lambda^2 t^2}.$$

This system has increasing hazard rate starting at zero when the system is new at $t = 0$, but over a long period of time the hazard rate approaches λ, the hazard rate of individual units.

2.7 The logistic and log logistic distributions

The logistic distribution is continuous with location parameter v and scale parameter τ and probability density function given by

$$f(t) = \frac{\tau^{-1}\exp[(t-v)/\tau]}{\{1 + \exp[(t-v)/\tau]\}^2}, \quad -\infty < t < \infty.$$

It is very similar to the normal distribution, but with longer tails, and if we take this form for log T, then we obtain a distribution similar to the lognormal called the log logistic distribution with probability density function

$$f(t) = \frac{\kappa t^{\kappa-1}\rho^\kappa}{[1 + (t\rho)^\kappa]^2}, \quad t \geq 0,$$

where $\kappa = \tau^{-1}$ and $\rho = e^{-v}$. The advantage of this family compared to the lognormal is the relatively simple forms obtained for the reliability, density and hazard functions. If $\kappa > 1$ the hazard has a single maximum, and if $\kappa < 1$ the hazard is decreasing.

2.8 The Pareto distribution

The generalized Pareto distribution has the hazard function

$$h(t) = \alpha + \frac{\beta}{\gamma + t}, \quad t \geq 0,$$

which can be increasing or decreasing according to the values of the parameters, α, β and γ. The reliability function is given by

$$R(t) = \exp\left[-\int_0^t h(t)\mathrm{d}t\right]$$

For $\alpha = 0$,

$$R(t) = \exp[-\beta \log (\gamma + t) + \beta \log \gamma]$$

$$= \left(\frac{\gamma}{\gamma + t} \right)^{\beta}$$

The mean residual lifetime is

$$m(t) = \frac{\int_t^\infty R(x)\mathrm{d}x}{R(t)} = \frac{\gamma + t}{\beta - 1}, \ \beta > 1.$$

An example of a situation giving rise to a Pareto model is one where failure time may be exponentially distributed with rate parameter not fixed but instead a random variable with a gamma distribution, say scale parameter β and shape parameter α:

$$f(t;\lambda) = \lambda e^{-\lambda t}, \qquad g(\lambda) = \frac{\beta(\beta\lambda)^{\alpha - 1} e^{-\beta\lambda}}{\Gamma(\alpha)}.$$

We can think of $f(t;\lambda)$ as the distribution of t conditional on λ. The joint distribution of t and λ, $f(t,\lambda)$ is given by $f(t;\lambda)g(\lambda)$ (this is analogous to (1.4)), and the unconditional distribution of t given by integrating $f(t,\lambda)$ over all λ. (Further examples of this type of mixing of distributions can be found in Chapter 10.)
So,

$$f(t) = \int_0^\infty f(t, \lambda)\mathrm{d}\lambda$$

$$= \int_0^\infty \lambda e^{-\lambda t} \frac{\beta(\beta\lambda)^{\alpha - 1} e^{-\beta\lambda}}{\Gamma(\alpha)} \mathrm{d}\lambda$$

$$= \beta^{\alpha} \int_0^\infty \frac{\lambda^{\alpha} e^{-(\beta + t)\lambda}}{\Gamma(\alpha)} \mathrm{d}\lambda.$$

Using $\Gamma(\alpha + 1) = \alpha\Gamma(\alpha)$ and the property that a gamma distribution integrates to one over $(0, \infty)$, we now write

$$f(t) = \frac{\alpha\beta^{\alpha}}{(\beta+t)^{\alpha+1}}\int_{0}^{\infty}\frac{(\beta+t)^{\alpha+1}\lambda^{\alpha}e^{-(\beta+t)\lambda}}{\Gamma(\alpha+1)}\,d\lambda$$

$$= \frac{\alpha\beta^{\alpha}}{(\beta+t)^{\alpha+1}}$$

since the function under the integral sign is equivalent to a gamma distribution with parameters $\beta + t$ and $\alpha + 1$.

To demonstrate that this is the probability density function of a Pareto distribution, we find

$$R(t) = \int_{t}^{\infty} f(x)\,dx = [-\beta^{\alpha}(\beta+x)^{-\alpha}]_{t}^{\infty}$$

$$= \left(\frac{\beta}{\beta+t}\right)^{\alpha}$$

and hence

$$h(t) = \frac{\alpha}{\beta+t}.$$

2.9   Order statistics and extreme value distributions

If the random variable X has probability density function $f(x)$ and distribution function $F(x)$, then a random sample from this distribution $X_1, X_2, ..., X_n$ rearranged in order of magnitude and denoted

$$X_{(1)} \le X_{(2)} \le ... \le X_{(n)}$$

gives the *order statistics* of the sample.

The joint probability density function of $X_{(1)}, ..., X_{(r)}$ is

$$\frac{n!}{(n-r)!}\left[\prod_{i=1}^{r} f(x_{(i)})\right][1 - F(x_{(r)})]^{n-r}.$$

The probability density function of $X_{(i)}$ $(1 \le i \le n)$ is

$$\frac{n!}{(i-1)!(n-i)!}F(x_{(i)})^{i-1}f(x_{(i)})[1-F(x_{(i)})]^{n-i}.$$

Moments of order statistics are not in general expressed very simply, but those for the uniform and exponential distributions are easily formulated and are useful since problems can often be transformed to problems for these distributions.

Example 2.5: If X has distribution function $F(x)$, then the random variables $F(X_{(i)})$, $i = 1, ..., n$ are order statistics for a sample of size n from the uniform distribution on [0, 1].

There are a number of *asymptotic* results associated with order statistics and these may be applied in various ways, in for example 'goodness-of-fit' tests. It can be shown, for example, that $X_{(i)}$ ($1 < i < n$), is *asymptotically* normal as $n \to \infty$, that is, the larger the number of order statistics, the more closely $X_{(i)}$ follows a normal distribution. Results of particular interest in reliability are those associated with the extreme order statistics $X_{(1)}$ and $X_{(n)}$.

The *central limit theorem* says that the sum of IID random variables has an asymptotic normal distribution, whatever the original probability distribution. In a similar way the largest (or smallest) of a set of IID random variables has (under mild conditions) a limiting distribution which takes one of three forms, referred to as Type I, II, III for maxima or minima. Leadbetter *et al.* (1983) give a comprehensive account of extreme value distributions and their applicability.

The Type I distribution for minima is the Gumbel distribution and the Type III distribution for minima is the Weibull distribution. These distributions are therefore often appropriate where the cause of failure is the weakest of a number of components. For example, material failure is often attributable to the weakest of a population of 'flaws' distributed throughout the material. If $\{X_i\}$ are the strengths of the flaws then the strength of the material is given by $X_{(1)}$.

An experiment in which n similar items are put on test at the same time and tested until failure will run as long as there is still one working item. In other words, if $\{X_i\}$ are the times to failure of the items, the length of the experiment will be given by $X_{(n)}$. Given n sufficiently large, the time until all items have failed may be modelled by an extreme value distribution. It may also be of interest to know the *total time on test* (see (9.1)) which here would be given by the sum of the IID lifetimes and would by the central limit theorem have approximately a normal distribution.

CHAPTER 3

Model Selection

3.1 Introduction

The choice of model for a given process may be governed by knowledge of the physical properties of the process. For example, in material strength the weakest-link property may be thought applicable and therefore a probability distribution with this feature would be appropriate. Past experience may also be a guide. If similar experiments have yielded results conforming to a particular distribution, that may be the first idea to pursue. In the absence of any pointers, some graphical exploratory data analysis is required to investigate such aspects as the behaviour of the empirical reliability or hazard functions, which may then suggest a plausible family of distributions. Whichever way is used to make the initial selection of a model, the fitting of that model must always be complemented by graphical and other means to assess how well the model fits the data (Chapter 4).

The named theoretical model, Weibull, exponential or some other well-defined form, can be thought of as specifying the model 'family'. This family will involve parameters which are usually unknown and require estimation from the available data. The parameter specification identifies the particular 'member' of the family being fitted to the data. Methods based on the assumption of a model family are called *parametric*. In order to make this initial assumption we need an estimate of the underlying probability distribution.

3.2 Non-parametric estimation of $R(t)$ and $h(t)$

Estimates of the reliability and hazard functions based purely on the data are referred to as *empirical*. Plotting techniques in exploratory data analysis designed to aid model selection are often based on these functions.

Consider a set of observations t_1, t_2, ..., t_n . The order statistics (Section 2.9) are given by rearranging the observations in ascending order of magnitude, $t_{(1)} \le t_{(2)} \le ... \le t_{(n)}$, where the subscript (i) has no particular relationship with subscript i. The estimated $R(t_{(i)})$ is related to the observed proportion of observations greater than $t_{(i)}$, say $1 - p_i$, where p_i is the observed proportion of observations less than or equal to $t_{(i)}$. An estimate of the reliability function is denoted by $\hat{R}(t)$. Setting $\hat{R}(t)$ equal to $1 - p_i$ results in a step function with jumps at the $t_{(i)}$.

There are a number of formulae for p_i for example

$$\frac{i-1}{n}, \frac{i}{n+1}, \frac{i-0.3}{n+0.4}, \frac{i-3/8}{n+1/4}, \frac{i-0.44}{n+0.12}, \frac{i-0.5}{n}.$$

The last is the Hazen formula and is in general the least biased for larger samples, say $n > 20$. For small samples $(i - 0.3)/(n + 0.4)$ is favoured. All but the first formula shown above are cases of the general formula $(i - \alpha)/(n - 2\alpha + 1)$. A discussion of the properties of these estimators can be found in Cunnane (1978).

Such methods of estimation which do not make any assumptions about the underlying distribution are called *non-parametric*.

The cumulative hazard function is given by $H(t) = - \log R(t)$ so it is natural to estimate $H(t)$ by

$$\hat{H}(t) = -\log \hat{R}(t).$$

From a plot of $\hat{H}(t)$ it is possible to assess whether the hazard function is increasing, decreasing, or constant. A linear plot implies a constant hazard function, a convex plot an increasing hazard, and a concave plot a decreasing hazard.

Direct estimation of the hazard function is given by

$$\hat{h}(t) = \frac{1}{(t_{(i+1)} - t_{(i)})(n - i + 1 - \alpha)},$$

where α takes the same value as used to estimate $R(t)$.

The ordinate heights of the density function are then given by

$$\hat{f}(t) = \frac{\hat{h}(t)}{\hat{R}(t)}.$$

With small data sets, or data collected at widely varying intervals, it may be necessary to employ some kind of 'smoothing' technique. It is wise to be aware that in small samples one spurious observation has considerable influence.

Example 3.1: The following table shows data from Kapur and Lamberson (1977) concerning observed kilocycles to failure for eight springs, in order of magnitude, and the resulting estimates of the reliability and hazard functions. For the purpose of smoothing, the fourth and fifth intervals have been combined and considered as a single interval in the hazard estimation. Figures 3.1 and 3.2 show plots of the estimated $R(t)$ and $h(t)$ with suitable functional forms.

i	$t_{(i)}$	p_i	$\hat{R}(t)$	$t_{(i+1)}-t_{(i)}$	$\hat{h}(t)$
1	190	0.083	0.917	55	0.0024
2	245	0.202	0.798	20	0.0075
3	265	0.321	0.679	35	0.0050
4	300	0.440	0.560	20	0.0171
5	320	0.560	0.440	5	
6	325	0.679	0.321	45	0.0082
7	370	0.798	0.202	30	0.0198
8	400	0.917	0.083		

3.3   Censoring

Lifetime data can very often be 'incomplete' in the sense that some observations may not be known exactly. A test on a number of units may have been stopped before all units have failed, so all that is known about some units is that they had a life exceeding the length of the test. Such observations are referred to as *right-censored* and are of the form $t_i > t_c$. It is also possible that some units may have failed prior to some initial time, that is, $t_i < t_c$ is recorded. This is called a *left-censored* observation. Units which fail at some unknown point between two observation times are said to be *interval-censored*. In practice, data which we classify as uncensored are in fact interval censored due to the rounding accuracy of nearly all measurement.

Censoring may arise for a variety of reasons, by design or otherwise. It is important to be aware of its existence and not to ignore any observation simply because a precise value is not known.

Two typical life tests are now described.

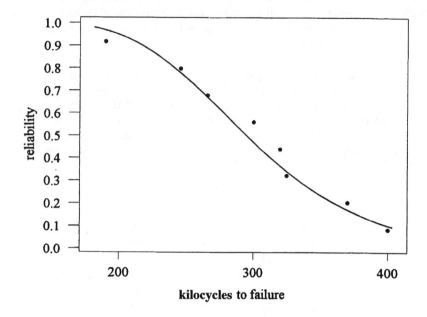

Figure 3.1. Reliability function for Example 3.1

(i)   A sample of n units are placed on test for limited periods
      of observation, C_1, C_2, ..., C_n , where possibly the C_i are
      all equal, but not necessarily. All observations are of the
      form $t_i = c \le C_i$ (uncensored) or $t_i > C_i$ forming a *Type I
      censored sample*.

(ii) A sample of n units are placed on test and the test is
      terminated when r of the units have failed. Again the
      observations are either uncensored or right-censored, but
      form what is termed a *Type II censored sample*.

With Type II censoring the number of uncensored lifetimes is fixed,
whereas for Type I censoring it is random.

Statistical analysis can be modified to take account of censored
data, provided choices concerning the number of failures or length of
test are effectively random. Censoring may occur due to accidental
damage, or units may be deliberately removed for inspection. In the
latter case units must be selected at random, and not because it is
thought they are about to fail.

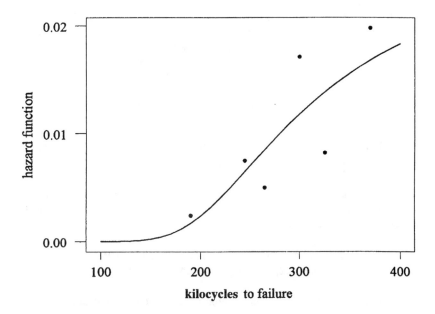

Figure 3.2. Hazard function for Example 3.1

3.4 Kaplan–Meier estimator

With Type II censored data, the most commonly used estimate for the reliability function is the *product-limit* or *Kaplan–Meier* estimate. It is defined as follows. Let observations be made at times $t_{(1)} < t_{(2)} < t_{(3)} < \ldots < t_{(k)}$, $k \leq n$, where n is the number of units on test. Let d_j be the number of failures at time $t_{(j)}$, and n_j the number of units 'at risk' at $t_{(j)}$, that is, the number of units unfailed and uncensored at $t_{(j)}$. The product-limit estimate of $R(t)$ is given by

$$
\hat{R}(t) = \frac{n_1 - d_1}{n_1} \cdot \frac{n_2 - d_2}{n_2} \cdot \frac{n_3 - d_3}{n_3} \ldots \frac{n_j - d_j}{n_j}, \; [j : (t_{(j)} < t, \, t_{(j+1)} > t)]
$$
$$
= \prod_{j \, : \, t_{(j)} < t} \frac{n_j - d_j}{n_j}.
$$

(3.1)

When all the data are uncensored this yields $\hat{R}(t_{(i)} + 0) = (n - i)/n = 1 - i/n$. This is a 'step' function with jumps just after the $t_{(i)}$. The time point just after $t_{(i)}$ is indicated by $t_{(i)} + 0$. The generally preferred

$\hat{R}(t)$ is $1 - (i - 0.5)/n$ which is $(1/2)\{1 - (i-1)/n + 1 - i/n\}$: This is yielded by 'smoothing' the product-limit estimate, taking the average of the estimated function before and after the observed times of failure. This is particularly important with small data sets. So it is best practice to produce plotting positions from

$$\hat{R}(t) = \frac{1}{2}\{\hat{R}(t_{(i-1)} + 0) + \hat{R}(t_{(i)} + 0)\}. \tag{3.2}$$

It should be noted that if the largest observed time is a censoring time, then $\hat{R}(t)$ is undefined beyond this point.

Example 3.2: The following are data from Lawless (1982) concerning remission times related to a drug for treating leukaemia. The starred observations are right-censored.

6, 6, 6, 6*, 7, 9*, 10, 10*,11*, 13, 16, 17*, 19*,
20*, 22, 23, 25*, 32*, 32*, 34*, 35*.

The following table can be constructed from these observations:

$t_{(j)}$	n_j	d_j	$\dfrac{n_j - d_j}{n_j}$	$\hat{R}(t_{(j)} + 0)$	$\hat{R}(t_{(j-1)} + 0)$	$\hat{R}(t) = \frac{1}{2}[\hat{R}_{(j-1)} + \hat{R}_{(j)}]$
6	21	3	18/21	0.857	1.000	0.929
7	17	1	16/17	0.807	0.857	0.832
10	15	1	14/15	0.753	0.807	0.780
13	12	1	11/12	0.690	0.753	0.722
16	11	1	10/11	0.627	0.690	0.659
22	7	1	6/7	0.538	0.627	0.583
23	6	1	5/6	0.448	0.538	0.493

It is noted that the reliability estimates only extend to the time of the largest uncensored observation.

Example 3.3: The effect of censoring is illustrated in this example. The cumulative hazard function, $H(t)$ will be estimated for the data, (i) treating all data as uncensored, and (ii) treating the observations in brackets as right-censored.

(5.8) 7.7 7.8 15.5 28.5 (38.8) 48.8 60.3 77.2 (173.8)

Given that $H(t) = -\log[R(t)]$ it is reasonable to estimate $H(t)$ by $-\log[\hat{R}(t)]$. We obtain the following tables:

(i)

$t_{(i)}$	$(i-0.5)/n$	$\hat{R}(t)$	$\hat{H}(t)$
5.8	0.05	0.95	0.051
7.7	0.15	0.85	0.163
7.8	0.25	0.75	0.287
15.5	0.35	0.65	0.431
28.5	0.45	0.55	0.598
38.8	0.55	0.45	0.798
48.8	0.65	0.35	1.050
60.3	0.75	0.25	1.386
77.2	0.85	0.15	1.897
173.8	0.95	0.05	2.996

(ii)

$t_{(i)}$	n_j	d_j	$\dfrac{n_j - d_j}{n_j}$	$\hat{R}_{(j-1)}$	$\dfrac{1}{2}[\hat{R}_{(j-1)} + \hat{R}_{(j)}]$	$\hat{H}$
7.7	9	1	0.889	1.000	0.944	0.057
7.8	8	1	0.875	0.889	0.833	0.182
15.5	7	1	0.857	0.778	0.722	0.325
28.5	6	1	0.833	0.667	0.611	0.492
48.8	4	1	0.750	0.556	0.486	0.721
60.3	3	1	0.667	0.417	0.347	1.058
77.2	2	1	0.500	0.278	0.208	1.569
				0.139		

Figure 3.3 shows these two estimated cumulative hazard functions plotted on the same scale. The hazard rate is characterized by the behaviour of the gradient of $H(t)$. In case (i) the hazard rate looks to be decreasing, but in case (ii) the impression is of an increasing hazard rate. Due to the censoring there is a smaller range over which is possible to estimate $H(t)$, and uncertainty over the true failure times of the censored items leads to higher influence of the uncensored observations.

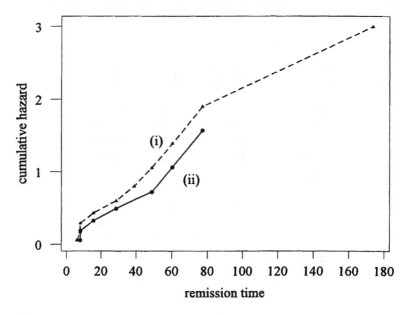

Figure 3.3. Cumulative hazard functions for Example 3.3

3.5   Graphical methods

Graphical methods form an important part of any statistically based analysis. Raw data can be examined in an intelligent way to get a feel for relevant factors and to point the way to suitable analyses. Here a group of plots will be considered which establish whether or not a particular model family is suitable for the data. Further plots will be considered in Chapter 4 which are appropriate after the modelling stage in order to check certain assumptions that may have been made.

The most useful plots are those which under the hypothesized model yield a straight line. A first assessment may then be given by eye, though it may still be the case that several hypothesized models appear plausible.

3.6   Straight line fitting

The best straight line through a set of points is commonly based on a 'least squares' principle (see the Appendix). This sets out to minimize a function of the distances of the points from the line. A straight line can be formulated as

$$y = mx + c, \tag{3.3}$$

where m is the gradient of the line, y relative to x, and c is the intercept of the line with the y axis. Straight line fitting is called *regression* in statistical modelling; y is termed the *response variable* and x the *explanatory variable*. In the context of probability plotting, x is a function of the observed random variable, t say, which may be strength or time for example, and y is a function of the estimated reliability function of t. Two questions are of interest. How well does the straight line represent the points? And what do the estimated values of m and c tell us about the chosen model for the distribution of t?

An answer to the first question is partly given purely by eye. The presence of some pattern to the way in which the points are scattered about the fitted line may indicate that the plot is not appropriate. A further measure of fit, useful in comparing plots, is the *R-squared* value. This is a function of the sums of squares of the distances between the points and the line, measured in the y direction. R-squared is given as a decimal/percentage in [0, 1] and a value close to one is the objective. This will be discussed further in the examples that follow.

3.7   Weibull plotting

In many cases a simple line plot can be deduced from the algebraic form of the model reliability function. A Weibull plot is obtained as follows:

$$R(t) = \exp[\ -(t/\alpha)^\beta\ ]$$

$$\log R(t) = -(t/\alpha)^\beta$$

$$\log[\ -\log R(t)\ ] = \beta \log t - \beta \log \alpha$$

If a Weibull distribution fits the data, then a plot of $y = \log[-\log \hat{R}(t_{(i)})]$ against $x = \log t_{(i)}$ will yield points lying approximately on a straight line. The $\hat{R}$ values are best calculated using the 'smoothed' Kaplan–Meier procedure (Section 3.4). Graphical estimates of α and β may be obtained by fitting the best straight line through the points and calculating the slope, m, and intercept, c.

The slope gives an estimate of β and the intercept yields $-\hat{\beta} \log \hat{\alpha}$, so

$$\hat{\alpha} = \exp\left(\frac{-\text{intercept}}{\hat{\beta}}\right). \tag{3.4}$$

This is equivalent to deducing $\hat{\alpha}$ from the point on the line where $\log[-\log \hat{R}] = 0$, at which $\log t = \log \alpha$.

Statistical packages will often do at least part of this process but care needs to be taken where censoring is present.

Example 3.4: Mann and Fertig (1973) give the following data for a life test of a sample of 13 of a particular aircraft component where the test was terminated after the tenth failure.

$$0.22 \quad 0.50 \quad 0.88 \quad 1.00 \quad 1.32 \quad 1.33 \quad 1.54 \quad 1.76 \quad 2.50 \quad 3.00$$

There are three right-censored observations at time 3.00. This must be allowed for in the calculation of $\hat{R}(t)$ and the plotting positions. Figure 3.4 shows the resulting Weibull plot with best fit straight line fitted by regression in MINITAB. The points lie fairly close to the line and there is no marked non-linearity. A Weibull model therefore seems appropriate.

The slope of the fitted line is 1.39, and this is an estimate of the shape parameter β. From (3.4) $\alpha = \exp(-(-1.14)/1.39) = 2.27$. 'R-squared' is the measure of fit referred to in Section 3.6, and at 0.98 indicates a high degree of fit. This is, however, principally a measure of how close the points are to the fitted line and should not be used without visual inspection of the graph as points can lie close to a line even when in a curve around the line.

Where lifetimes are thought to follow a Gumbel distribution (of Sections 2.4 and 2.9), a plot using the same procedure as above, but with the $t_{(i)}$ coordinates replaced by values of $\log t_{(i)}$, will similarly be approximately linear.

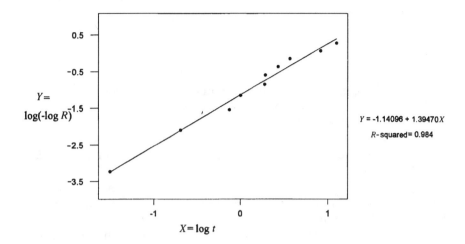

Figure 3.4. Weibull plot for Example 3.4

3.8   Normal plotting

The standard normal distribution was defined in Section 2.5. This is the normal distribution with mean 0 and variance 1.

Let a general normal variable be t with mean μ and variance σ^2; then the standardized score,

$$z = \frac{t - \mu}{\sigma}, \tag{3.5}$$

has a standard normal distribution. Standardized scores and their corresponding distribution function values can be found from normal tables. The distribution function is $\Phi(z)$ and is estimated by $1 - \hat{R}(t)$. For an ordered observation $t_{(i)}$ the corresponding $z_{(i)}$ under the assumption of a normal model is given by

$$\Phi^{-1}(1 - \hat{R}(t_{(i)})).$$

The quantity $\Phi^{-1}(.)$ is sometimes referred to as a *normal score*.

If the model assumption is reasonable relationship (3.5) will hold. Rearranging gives

$$t = \sigma z + \mu,$$

which is linear in z. A plot of $t_{(i)}$ against $z_{(i)}$ therefore provides a simple model-checking graph. Estimates of σ and μ can be given by the slope and intercept parameters but there are better, simple ways if the data are uncensored (see Section 4.3).

Plots which yield a straight line and rough estimates of scale and location parameters from the slope and intercept are called *quantile-quantile* (QQ) plots.

Again there is an analogous plot if log $t_{(i)}$ replaces $t_{(i)}$. This will yield a straight line if the data are from a lognormal distribution. It is important to remember that the parameters μ and σ refer to the resulting normal distribution of the log $t_{(i)}$. The mean and variance of the $t_{(i)}$ are functions of μ and σ but are not simply anti-logarithmic functions (see Example 7.2). For example,

$$\log(\text{mean } t) \neq \text{mean}(\log t).$$

Example 3.5: The following data are from Lieblein and Zelen (1956) and have been subject to much analysis over the years. The ordered observations are the number of millions of revolutions to failure for each of 23 ball bearings.

17.88	28.92	33.00	41.52	41.12	45.60	48.40	51.84
51.96	54.12	55.56	67.80	68.64	68.64	68.88	84.12
93.12	98.64	105.12	105.84	127.92	128.04	173.40	

These data have mean 72.22, so the log of the mean is 4.28. However, if the data are first logged the mean is then 4.15.

Example 3.6: For the data of Example 3.4 the following shows the calculation of the coordinates upon which a normal plot is based.

$t_{(i)}$	$\log(t_{(i)})$	$\hat{R}(t)$	$z = \Phi^{-1}(1 - \hat{R}(t))$
0.22	−1.51	0.962	−1.769
0.50	−0.693	0.885	−1.198
0.88	−0.128	0.808	−0.7869
1.00	0.0	0.721	−0.615

$t_{(i)}$	$\log(t_{(i)})$	$\hat{R}(t)$	$z = \Phi^{-1}(1 - \hat{R}(t))$
1.32	0.278	0.654	−0.396
1.33	0.285	0.577	−0.194
1.54	0.432	0.500	0.0
1.76	0.565	0.423	0.194
2.50	0.916	0.346	0.396
3.00	1.099	0.269	0.615

Figure 3.5 shows a normal plot based on the logged data. The slope and intercept of the fitted line on the lognormal plot give estimates of the mean and standard deviation of the population of log times, that is $\hat{\mu} = 0.514$ and $\hat{\sigma} = 1.017$. Compared to the Weibull plot of Figure 3.4, the fit is less good with some curvature of the points about the line. The 'S' shape of the plotted points is indicative of a distinct difference between the data and the proposed model. The R-squared value is 0.966 compared to 0.984 in the Weibull case, again giving more weight to the Weibull model.

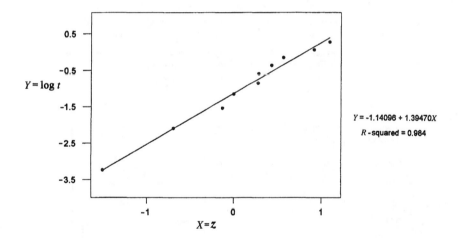

Figure 3.5. Lognormal plot for Example 3.4

3.9   Other model family plots

A number of model families may be investigated in a similar way to those above, based either on the algebraic form of the reliability function or on suitable exact or approximate use of the normal distribution function. A summary of the most useful plots is given in Table 3.1.

model	y	x
exponential	$\log \hat{R}(t)$	t
Pareto	$\log \hat{R}(t)$	$\log t$
Gumbel	$\log[-\log \hat{R}(t)]$	t
Weibull	$\log[-\log \hat{R}(t)]$	$\log t$
normal	$\Phi^{-1}(1 - \hat{R}(t))$	t
lognormal	$\Phi^{-1}(1 - \hat{R}(t))$	$\log t$
gamma	$\Phi^{-1}(1 - \hat{R}(t))$	$\sqrt{t}$
logistic	$\log[(1 - \hat{R}(t))/\hat{R}(t)]$	t
log logistic	$\log[(1 - \hat{R}(t))/\hat{R}(t)]$	$\log t$

Table 3.1. Model family plots

Example 3.7: Figures 3.6–3.8 show plots for the data of Example 3.5 in relation to the Weibull, lognormal and gamma models. The Weibull and lognormal plots have been produced by MINITAB routines for uncensored data. The slight curvature in the scatter of points about the line in Figure 3.5 tends to give more favour to the lognormal and gamma models, but there is little help in distinguishing between the latter.

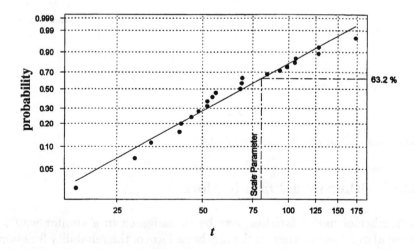

Figure 3.6. Weibull plot for Example 3.7

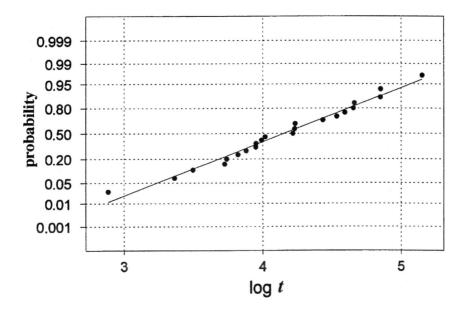

Figure 3.7. Lognormal plot for Example 3.7

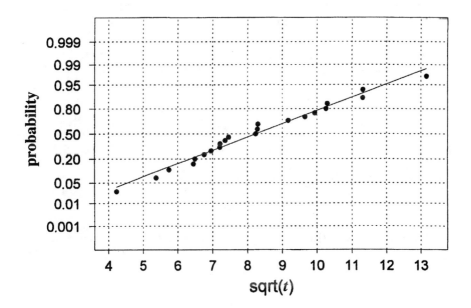

Figure 3.8. Gamma plot for Example 3.7

3.10   Comparison of distributions

One of the most difficult problems is differentiating between two or
more models which appear equally plausible for a given set of data.
The choice is important because, for example, estimated quantiles, say
in the lower tail, may differ considerably. Often there is insufficient
data to choose between different forms purely by empirical analysis.
The choice may then be made on other grounds, for example the
physical properties of the process, technical convenience, or the behaviour of the distributions in their tails.

For uncensored data there is more choice in the assessment techniques available. A useful exploratory technique based on sample
moments is described in Cox and Oakes (1984). For a random variable
X with distribution mean μ and standard deviation σ, the *coefficient
of variation* is given by $\gamma = \sigma/\mu$ and the *standardized third moment* is
given by $\mu_3^* = \mu_3/\sigma^3$, where

$$\mu_3 = E[(X - \mu)^3].$$

Distribution families which have a single shape parameter have a
particular $\mu_3^* = f(\gamma)$. It can be shown, for example, that for the gamma
distribution $\mu_3^* = 2\gamma$. The proximity to a particular curve of the point
representing the equivalent sample measures may indicate a likely
choice of model.

For data $t_1, t_2, \ldots, t_n$ the coefficient of variation is estimated by $s/\bar{t}$
where $\bar{t}$ and s are respectively the *sample mean* and *sample standard
deviation*, defined by

$$\bar{t} = \frac{1}{n}\sum_{i=1}^{n} t_i \qquad\qquad (3.6)$$

$$s^2 = \frac{1}{n-1}\sum_{i=1}^{n} (t_i - \bar{t})^2. \qquad\qquad (3.7)$$

The value of μ_3^* is estimated by the *sample standardized third moment*,

$$\hat{\mu}_3^* = \frac{1}{n}\sum_{i=1}^{n}\left(\frac{t - \bar{t}}{s}\right)^3.$$

The data of Example 3.5 give sample parameters as follows:

$$\bar{t} = 72.22, \quad s = 37.49, \quad \hat{\gamma} = 0.52, \quad \hat{\mu}_3^* = 0.88.$$

Figure 3.9 shows a diagram similar to that of Cox and Oakes (1984, p. 27) with the point (0.52, 0.88) corresponding to these data. There is no clear distinction between the suitability of the Weibull, gamma and lognormal distributions as models of the lifetime of these ball bearings, but perhaps an indication that the gamma or Weibull distribution may be the preferred choice. Distinguishing between these three models is in fact a very common problem in lifetime analysis.

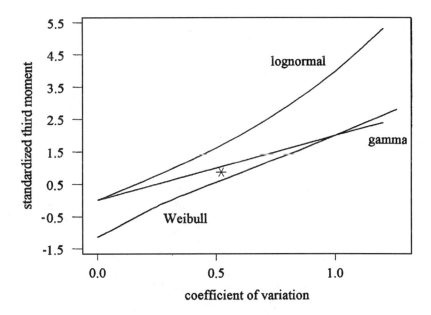

Figure 3.9. Cox and Oakes moments plot

Criteria for model fit will be discussed in the next chapter.

CHAPTER 4

Model Fitting

4.1 Parameter estimation

The chosen parametric model is fitted to the data by estimating the unknown parameters, θ. The data are observations which provide us with *information* about the process being observed. The amount of information can be thought of as fixed in a sense, though how much we can reap from that information may depend on the efficiency of the method used to extract the information. Different sets of observations, even when collected under similar conditions, yield different information. Good methods of estimation will have the properties of accuracy (low bias), efficiency (low variability) and consistency. However, the basic fact is that there is only so much information to be got from the data. More data will yield more information. If more unknown parameters are put into the model, the more thinly the information will be spread and thus there will be more uncertainty in the parameter estimation.

The choice of estimation method may in practice be governed by considerations such as ease of use rather than statistical merit. There are, for example, potentially severe limitations to some methods in the presence of censored data.

The estimation process requires answers to the following questions.

(i)   Are the lifetime data censored or uncensored?
(ii) Is there a model known, or assumed, to fit the population from which the data arise?
(iii) What is the purpose of the estimation?

4.2 The variance of estimators

Point estimates on their own do not tell us how close we might be to the true value of the parameter concerned. Further, different samples from the same population will give different estimates of the same parameter. However, we can determine the degree to which these estimates might vary and hence give a range of plausible values for the parameter. This is called an *interval estimate* and provides more information than a point estimate.

The variability in parameter estimates is characterized by the *standard error* (s.e.), given by the square root of the variance of the estimator. If θ^* is an estimator for θ then the variance of θ^* is given by

$$V(\theta^*) = E[\theta^* - E(\theta^*)]^2.$$

Another measure of interest is the *mean squared error* (m.s.e.), defined by $E[\theta^* - \theta]^2$.

When the estimator θ^* is *unbiased*, $E[\theta^*] = \theta$ and therefore the mean squared error and the variance are the same. Otherwise the mean squared error is given by the variance plus a contribution which is the square of the *bias*, $E[\theta^*] - \theta$:

$$E[\theta^* - \theta]^2 = E[\theta^* - E(\theta^*) + E(\theta^*) - \theta]^2$$
$$= E[\theta^* - \theta]^2 + E\{2[\theta^* - E(\theta^*)][E(\theta^*) - \theta]\} + E[E(\theta^*) - \theta]^2.$$

The middle term is zero because $[E(\theta^*) - \theta]$ is a constant and $E[\theta^* - E(\theta^*)]$ is zero. Therefore

$$\text{m.s.e.}(\theta^*) = V(\theta^*) + \text{bias}^2. \tag{4.1}$$

We can think of the mean squared error as having two components, precision/efficiency and accuracy. High precision, that is low variability, is usually considered to be the most important. Bias can often be quantified and a suitable correction made. The properties of the mean squared error are illustrated well in Chatfield (1983, p. 119).

4.3   Confidence interval estimates

Confidence limits for estimated parameters, or functions of estimated parameters, are based on the point estimates combined with functions of the standard errors.

In general, an estimate $\hat{\theta}$ is given by some $f(\mathbf{X})$ where $\mathbf{X} = \{X_1, X_2, X_3...\}$ is a random sample from the population considered. Many estimates are approximately normally distributed, provided the sample size is large, and confidence bounds are of the form

$$E[f(\mathbf{X})] \pm z_c \sqrt{V[f(\mathbf{X})]}, \qquad (4.2)$$

where z_c is a percentage point from the standard normal distribution reflecting the degree of belief in the limits. For example, bounds cited with 95% confidence are based on the 2.5% and 97.5% points of the standard normal distribution. It is often the case in reliability estimates that only a lower bound is of interest. Frequently the upper bound is 1, which is not very informative. In this case a 95% lower bound would use the 5% point of the distribution of the estimate.

Consider the question of whether data conforming to a Weibull distribution might be adequately represented by the simpler exponential model. The latter is the case $\beta = 1$. An assessment of the suitability of the exponential model may be made by using the regression information provided by the statistical routine in order to put an approximate confidence interval on the slope parameter. If this confidence interval contains the value 1, then there is insufficient evidence to reject the exponential model in favour of the Weibull model with $\beta \neq 1$.

Example 4.1: Table 4.1 shows the regression output provided by MINITAB for the Weibull plot of Example 3.4. The straight line fit is good (Figure 3.4) and a Weibull model appropriate.

Table 4.1 shows for each regression coefficient its standard error, *t-ratio* and *p-value*. The high *t*-ratios and very small *p*-values simply indicate that the coefficients are significantly different from zero, which is what we would expect. The standard errors are of most interest, particularly for the estimate of the slope. Taking $\hat{\beta}$ to be approximately normally distributed and using Equation (4.2), a 95% confidence interval estimate of the slope is given by

$$1.3947 \pm 1.96 \times 0.06368 = [1.27, 1.51].$$

There is thus reason to believe that a slope of 1 is not likely, that is, the Weibull model is a more appropriate choice than the exponential.

```
The regression equation is lg(-lgR) = -1.14 + 1.39 log t
Predictor          coeff        stdev     t-ratio      p
constant         -1.14096      0.04723    -24.16     0.000
log t             1.39470      0.06368     21.90     0.000

s = 0.1472     R-sq = 98.4%
```

Table 4.1. MINITAB regression output for Example 3.4

4.4   Maximum likelihood

Graphical estimation of parameters is often simple and effective, but is prone to disproportionate influence by unusual observations. The most popular method of estimation is *maximum likelihood,* due to its generally favourable properties. For uncensored samples, the likelihood function is defined as

$$L(\boldsymbol{\theta};\mathbf{t}) = \prod_{i=1}^{n} f(t_i ; \boldsymbol{\theta}) . \tag{4.3}$$

The likelihood is proportional to the probability of getting the observed data under the model type assumed. It is a function of the unknown parameters $\boldsymbol{\theta}$ and the estimation proceeds under the assumption that the data observed were the most likely data to occur.

Maximum likelihood estimates of parameters are those values which make the likelihood function as large as possible, that is, maximize the probability of the observed sample. For computational convenience the log likelihood is generally used. So (4.3) is transformed to give

$$\log L(\boldsymbol{\theta}; \mathbf{t}) = \sum_{i=1}^{n} \log f(t_i ; \boldsymbol{\theta}) .$$

The equations $\partial \log L/(\partial \theta_i) = 0$ have to be solved; this is frequently not possible analytically, and an iterative numerical method may be required. Where a graphical method of estimation is available, this can provide good initial values for the parameter estimates. Alternatively, it may be convenient to numerically maximize the likelihood function directly.

Censored observations make contributions to the likelihood which are functions of the reliability function. An observation which is right-censored at t_C contributes $R(t_C ; \boldsymbol{\theta})$; one which is left-censored at t_C contributes $1 - R(t_C ; \boldsymbol{\theta})$; finally, an interval-censored observation contributes $R(t_a ; \boldsymbol{\theta}) - R(t_b ; \boldsymbol{\theta})$ for $t_a < t_b$. For example, given data consisting of a set U of uncensored observations and a set C of right-censored observations, the likelihood function is

$$\prod_{i \in U} f(t_i; \boldsymbol{\theta}) \prod_{i \in C} R(t_i; \boldsymbol{\theta}) .$$

Example 4.2: The following data are believed to be observations from an exponential distribution with unknown parameter λ:

$$12^*, \ 13, \ 6, \ 10, \ 11, \ 7, \ 11, \ 7, \ 15, \ 8^*.$$

Starred values are right-censored.

The likelihood function is given by

$$\prod_U \lambda e^{-\lambda t_i} \prod_C e^{-\lambda t_i} .$$

The log likelihood is then

$$\sum_U (\log \lambda - \lambda t_i) + \sum_C -\lambda t_i = k \log \lambda - \sum_{i=1}^{n} \lambda t_i ,$$

where n is the total number of observations and k is the number of uncensored observations. Then

$$\frac{\partial \log L}{\partial \lambda} = \frac{k}{\lambda} - \sum_{i=1}^{n} t_i . \tag{4.4}$$

Setting this to zero yields

$$\hat{\lambda} = \frac{k}{T} ,$$

where $T = \sum_{i=1}^{n} t_i$ and is often referred to as the *total time on test*. When $k = n$, that is, all observations are uncensored, then $\hat{\lambda} = 1/\bar{t}$, where $\bar{t}$ is

the sample mean. This is an intuitive estimate of λ given that the mean
of the exponential distribution is $1/\lambda$.

For the above data, $T = 100$ and $k = 8$ giving $\hat{\lambda} = 0.08$. Figure 4.1 shows
the log likelihood function, which is typically of this form.

It should be emphasized that maximum likelihood estimation
identifies the member of a particular model family most appropriate
to the data, but does not on its own demonstrate the suitability of
the model family.

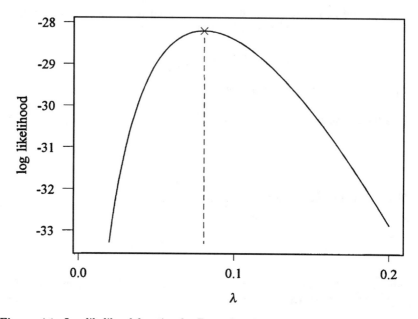

Figure 4.1. Log likelihood function for Example 4.2

Example 4.3: For the data of Example 3.4 graphical estimates of the
Weibull parameters were $\hat{\alpha} = 2.27$ and $\hat{\beta} = 1.39$. Because the fit here
is good, the graphical estimates of the model parameters are in close
agreement with the maximum likelihood estimates, which are $\hat{\alpha} = 2.27$
and $\hat{\beta} = 1.42$.

Example 4.4: Consider the normal distribution described in Section 2.5.
Suppose we wish to estimate the parameters μ and σ^2 using maximum
likelihood assuming a sample of uncensored observations. We can show
using (4.3) that

$$\hat{\mu} = \bar{t}$$

and

$$\hat{\sigma}^2 = \frac{1}{n}\sum_{i=1}^{n}(t_i - \bar{t})^2 .$$

The latter is an example of a maximum likelihood estimate which is biased. $E(\hat{\sigma}^2)$ is not σ^2 but $(1 - 1/n)\sigma^2$. However, this estimate has good properties as far as the normal distribution is concerned. In general it is custom to estimate variance using the unbiased estimate, s^2, given by (3.7). The estimate $\hat{\sigma}^2$ has bias but it has smaller mean squared error than s^2 (see Section 4.2). The difference between s^2 and $\hat{\sigma}^2$ decreases as the sample size increases.

Where normal data are censored graphical estimates are very useful because maximum likelihood estimation is complicated by contributions to the likelihood involving the non-analytic function Φ.

The construction of confidence intervals for the estimated parameters will involve the maximum likelihood estimates and their standard errors. Determination of the exact distribution of maximum likelihood estimators is only possible in certain special cases. Simulation will yield close approximations of distributional properties but common use is made of the asymptotic normality of maximum likelihood estimates.

The *observed information matrix* has entries $-\partial^2 \log L/\partial\theta_i\partial\theta_j$ evaluated at $\hat{\theta}$, and the inverse of this matrix is the estimated *covariance matrix* of $\hat{\theta}$. Confidence intervals for the θ_i are given by

$$\hat{\theta}_i \pm z_C \,(\text{s.e. } \hat{\theta}_i) \tag{4.5}$$

where z_C is the appropriate percentage point from the standard normal distribution.

Improved results can sometimes be obtained for small samples by reparameterizing the distribution. For example, we might write $\lambda_i = \log \theta_i$ and $\hat{\lambda}_i = \log \hat{\theta}_i$ may be more closely normal than $\hat{\theta}_i$.

Example 4.5: Continuing Example 4.2, the covariance matrix for a single parameter distribution is simply the reciprocal of minus the second derivative of the log likelihood. Differentiating (4.4) with respect to λ,

$$\frac{\partial^2 \log L}{\partial \lambda^2} = -\frac{k}{\lambda^2}.$$

Evaluating this derivative at $\lambda = \hat{\lambda}$ yields $V(\hat{\lambda}) = \hat{\lambda}^2/k$. If normality is assumed for $\hat{\lambda}$, then a 95% confidence interval for $\hat{\lambda}$ is given by

$$\hat{\lambda} \pm 1.96\frac{\hat{\lambda}}{\sqrt{k}} = 0.08 \pm 1.96\frac{0.08}{\sqrt{8}} = [0.0246, 0.1354].$$

This neatly demonstrates how censored observations contribute to the point estimation process but yield little information with regard to the precision of such estimates.

These parameter bounds may be translated into, for example, reliability bounds. In the case of the exponential distribution, a lower bound on reliability will correspond to the upper bound for λ. That is,

$$\exp\left\{-\left(\hat{\lambda} + 1.96\frac{\hat{\lambda}}{\sqrt{k}}\right)t\right\}. \tag{4.6}$$

We can use $\exp\{-\hat{\lambda}t\}$ as the maximum likelihood estimate of $R(t)$ because maximum likelihood estimates have a property known as *invariance*. This means that maximum likelihood estimates of functions of parameters are simply the same functions applied to the maximum likelihood estimates of the parameters. To put this concisely,

$$(\widehat{\phi(\theta)}) = \phi(\hat{\theta}).$$

When θ is a vector of parameters, putting confidence bounds on $\phi(\theta)$ is not necessarily straightforward, as it is in the case illustrated by (4.6). A method for estimating the variance of a function of several estimated parameters is shown in Chapter 7.

Comparison of maximized log likelihoods under different models may be used as an indication of model suitability. This has wide capability as it can be used with all kinds of censored data. Broadly speaking, the higher the log likelihood the better the fit.

Example 4.6: Crowder *et al.* (1991) examine the Lieblein and Zelen data of Example 3.5 and fit both Weibull and lognormal models. The plots in Figures 3.6 and 3.7 show no clear preference for either of the models. Crowder *et al.* show the maximized log likelihoods to be −18.24 for the Weibull model and −17.27 for the lognormal. The lognormal distribution would be considered the better fit as it has the largest likelihood. However, a formal test of whether one distribution fits significantly better than another is only straightforward if the models are from the same family. For example, we might test whether there is any real advantage in using a general Weibull distribution rather than the simpler exponential, which is a special case of a Weibull distribution. Formal testing is considered in more detail in Chapter 9. For the above data, Crowder *et al.* cite the work of Dumonceaux and Antle (1973) in discriminating between the Weibull and lognormal models and conclude that the difference in the likelihoods is non-significant.

4.5 Estimating quantiles

The warranty period, guaranteed or 'safe' life of a product or component is often of practical interest. A certain reliability is specified, say no more than 1% failure within the guaranteed lifetime, and the question is the value of that lifetime. We can best answer that question if we have a good idea of the lifetime model. Let the probability of failure within the guaranteed lifetime, t_p , be p. Then if the lifetime has reliability function $R(t)$ the value of t_p is given by solving

$$1 - R(t_p) = p.$$

The lifetime t_p is called a *quantile*. The quantiles of interest are those in the tails of the lifetime distribution, and this is where the estimation process may be the most difficult. By definition, these tail values do not occur very often and a typical sample of modest size will tend to consist of lifetimes in the region of the lifetime distribution which has the bulk of probability. It may well be the case that several models fit these data acceptably well, but estimation of quantiles outside the range of observed lifetimes will often yield quite different values under the different models.

Example 4.7: Following on from Example 4.6, these data will be used to estimate $t_{0.01}$ assuming Weibull, lognormal and gamma distributed lifetime.

(i)   Weibull. The maximum likelihood estimates of α and β are respectively 81.87 and 2.1. So $t_{0.01}$ is given by t in

$$0.01 = 1 - \exp\left[-\left(\frac{t}{81.87}\right)^{2.1}\right],$$

$$t = 81.87[-\log(0.99)]^{1/2.1} = 9.16.$$

(ii)   Lognormal. $\log t$ has a normal distribution with parameters μ and σ^2. The maximum likelihood estimates of μ and σ are respectively 4.15 and 0.522. The 1% point on the $\log t$ distribution is given by y in

$$0.01 = \Phi\left(\frac{y - 4.15}{0.522}\right)$$

So, from normal tables,

$$\frac{y - 4.15}{0.522} = -2.327$$

$$y = 2.935.$$

Therefore,

$$t = e^y = 18.83.$$

(iii)   Gamma. Recalling the form of the gamma distribution function from Section 2.6, the likelihood function is not at all user-friendly. Maximizing the function requires numerical routines to deal with $\Gamma(\alpha)$ as well as the optimization. Further, the reliability function does not have closed form, nor is it available in tabular form. The value of $t_{0.01}$ is given by t in

$$0.01 = \int_0^t \frac{\lambda(\lambda x)^{\alpha-1}e^{-\lambda x}}{\Gamma(\alpha)}dx.$$

This kind of calculation can be carried out using routines in the computational package MATLAB. There is also an extensive list of routines for a wide variety of applications in Press *et al.* (1992). This volume is for Fortran programmers but similar editions are available in Basic, Pascal and C. It turns out that here the maximum likelihood estimates of α and λ are 4.0247 and 0.0557, respectively. So

$$\frac{0.01\Gamma(4.0247)}{(0.0557)^{4.0247}} = \int_0^t x^{3.0247} e^{-0.0557x} dx .$$

Solving this equation iteratively yields $t = 12.9$.

To summarize, the Weibull, lognormal and gamma models yield respectively for $t_{0.01}$ the values 9.16, 18.83 and 12.9. There proves to be little real statistical evidence of a distinct preference for any one of these models, but the choice has considerable implication with regard to estimation of tail quantiles. For the purposes of specifying a 'safe life' the Weibull model would appear to be the choice with lowest risk. However, comparisons of likelihood functions and the graphical analyses shown in Chapter 3 indicate that this could be rather conservative. From a cost point of view there may be a considerable increase in cost as a result of this choice.

4.6   Estimation methods using sample moments

For cases where maximum likelihood is difficult computationally and where only uncensored data are available, a simple, fairly efficient method of estimation is provided by equating population and sample moments – the *method of moments*. The gamma distribution provides such an example. The model has two parameters, so two equations involving moments are required. The model mean and variance are respectively

$$\mu = \frac{\alpha}{\lambda}, \sigma^2 = \frac{\alpha}{\lambda^2}$$

giving

$$\frac{\mu}{\sigma^2} = \lambda .$$

So method of moments estimation yields

$$\hat{\lambda} = \frac{\bar{x}}{s^2}, \ \hat{\alpha} = \bar{x}\hat{\lambda} .$$

Example 4.8: Again taking the data of Example 3.5 and assuming a gamma model, the method of moments estimators of α and λ are

$$\hat{\lambda} = \frac{72.22}{37.49^2} = 0.0514, \ \hat{\alpha} = 72.22 \times 0.0514 = 3.71 \ .$$

These are lower than the maximum likelihood estimates in Example 4.7, but compare reasonably well.

The Weibull distribution does not lend itself to this approach very readily because the population moments are not of user-friendly form − see Section 2.3. However, reasonable estimates of the Gumbel distribution parameters may be obtained via sample moments and estimates of the Weibull parameters can be deduced from these. Recalling Section 2.4, if T has a Weibull distribution with parameters α and β, then $Y = \log T$ has a Gumbel distribution with parameters $\eta = 1/\beta$ and $\gamma = \log \alpha$. If for Weibull data the logs are calculated and then the mean, $\bar{y}$, and standard deviation, s, calculated, simple estimates of η and γ are given by

$$\hat{\eta} = \frac{\sqrt{6}}{\pi}s, \ \hat{\gamma} = \bar{y} + 0.5772\hat{\eta} \ , \qquad (4.7)$$

where 0.5772 is Euler's constant.

Example 4.9: For the data of Example 3.5, the logged data have mean 4.15 and sample standard deviation $s = 0.5334$. So estimates of the Weibull parameters may be given by

$$\hat{\beta} = \pi / (0.5334\sqrt{6}) = 2.40$$

and

$$\hat{\alpha} = \exp\left[4.15 + 0.5772 \times \frac{1}{2.40}\right] = 80.64 \ .$$

These estimates compare favourably with the maximum likelihood estimates, 2.1 and 81.87, respectively.

The estimates for η and γ are biased, but for large samples, the distribution of the estimates is approximately normal:

$$\hat{\eta} \sim N\left(\eta, \frac{1.1\eta^2}{n}\right),$$

$$\hat{\gamma} \sim N\left(\gamma, \frac{1.168\eta^2}{n}\right),$$

where n is the number of observations in the data set. Hence we can construct confidence interval estimates for the parameters.

Example 4.10: For the data of Example 3.5 Equations (4.7) yield

$$\hat{\eta} = 0.417, \ \hat{\gamma} = 4.39,$$

and 95% confidence interval estimates are given by

$$\hat{\eta} \pm 1.96 \sqrt{\frac{1.1}{n}} \hat{\eta} = [0.238, 0.596]$$

and

$$\hat{\gamma} \pm 1.96 \sqrt{\frac{1.168}{n}} \hat{\eta} = [4.206, 4.574].$$

These can then be translated into interval estimates for the Weibull parameters.

4.7   General probability plots

These plots, often called *PP plots*, are distribution-specific, in that parameter values have to be provided in a given model in order to calculate the plotted points. A specific distribution may then be compared to the empirical distribution function by plotting $\hat{F}(t)$ against the fitted $F(t)$, or equally $\hat{R}(t)$ against the fitted $R(t)$. A good fit will be implied by points closely approximating the straight line between (0, 0) and (1, 1). The essence of this plot is that a non-parametric estimate of the probability distribution is compared to the proposed parametric estimate. A good model fit will yield for each observed value of t two probability values in close agreement.

Example 4.11: Figure 4.2 shows for the Mann and Fertig data of Example 3.4 points with coordinates

$$\left(\hat{R}(t_{(i)}), \ \exp\left\{ -(t_{(i)}/\hat{\alpha})^{\beta} \right\} \right).$$

The first coordinate is the empirical reliability, as shown in Example 3.6, and the second coordinate is the fitted reliability using a Weibull model and maximum likelihood estimates for the parameters. These two values are close to each other across the observed $t_{(i)}$ and therefore the points sit close to the straight line linking (0,0) and (1,1). Also the points have no particular pattern in their departure from the line, so the model fit is good.

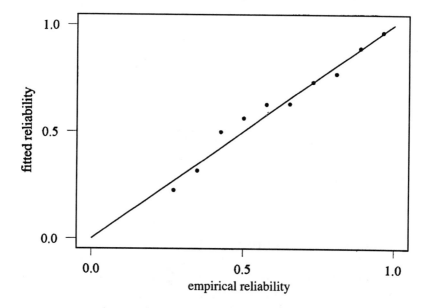

Figure 4.2. PP plot for Example 4.11

Example 4.12: Table 4.2 concerns the strength of carbon fibres, tension-tested in air. This is a large data set, 137 uncensored observations, relating to the breaking strength, in gigapascals, of fibres of approximate diameter 7.8 micrometres and length 5 millimetres. The strength distribution is the subject of Wolstenholme (1996).

The brittle nature of carbon would suggest a weakest-link property and a Weibull model would be a natural choice. Figure 4.3 shows a Weibull plot produced by MINITAB directly from the uncensored data set. A poor

Fibre	Strength	Diameter	Fibre	Strength	Diameter	Fibre	Strength	Diameter
1	3.30	7.91	47	4.52	7.09	93	3.40	8.40
2	4.17	7.29	48	4.60	7.12	94	3.28	8.43
3	4.19	6.59	49	5.60	7.14	95	3.02	9.13
4	3.64	8.45	50	4.38	7.28	96	3.06	9.10
5	2.73	9.01	51	4.38	7.42	97	3.20	8.55
6	4.47	6.82	52	4.35	7.42	98	3.96	7.92
7	3.29	7.43	53	3.94	7.43	99	3.73	8.07
8	3.55	8.19	54	4.17	7.43	100	3.70	8.13
9	3.03	9.01	55	4.34	7.44	101	4.04	7.92
10	6.41	6.52	56	4.30	7.73	102	3.83	7.94
11	5.16	6.92	57	4.94	7.80	103	3.92	7.98
12	4.92	7.01	58	3.92	7.84	104	4.78	8.05
13	3.01	7.73	59	3.90	7.85	105	4.10	7.81
14	4.01	8.49	60	3.98	7.90	106	4.06	7.81
15	5.09	7.79	61	3.85	7.92	107	3.68	8.18
16	4.65	6.92	62	3.80	7.92	108	3.15	8.18
17	4.57	7.26	63	3.82	7.92	109	4.44	7.36
18	3.48	8.23	64	3.72	7.92	110	4.47	7.41
19	3.05	9.03	65	3.70	8.05	111	4.31	7.42
20	3.60	7.73	66	3.72	8.06	112	4.18	7.71
21	4.60	6.94	67	3.71	8.08	113	3.82	7.40
22	4.38	7.42	68	3.68	8.10	114	3.81	7.76
23	3.50	7.74	69	3.53	8.34	115	4.01	7.40
24	4.43	7.09	70	3.35	8.40	116	4.12	6.69
25	4.58	6.97	71	3.28	8.42	117	3.50	7.41
26	4.76	6.92	72	3.30	8.43	118	3.63	7.76
27	4.64	6.93	73	3.23	8.45	119	3.69	7.29
28	2.65	9.42	74	3.10	8.84	120	4.20	7.70
29	5.03	6.92	75	2.94	9.10	121	4.93	6.97
30	5.15	6.95	76	3.80	7.76	122	4.58	7.23
31	3.35	7.92	77	3.73	7.43	123	4.09	7.76
32	3.62	7.91	78	3.69	7.45	124	4.17	7.81
33	4.04	7.92	79	3.42	7.69	125	3.69	8.21
34	3.06	9.01	80	4.16	7.79	126	4.20	7.69
35	4.55	8.09	81	3.98	7.89	127	4.26	7.70

continued

Table 4.2. Strength and diameter measurements for 137 carbon fibres tension-tested in air

Fibre	Strength	Diameter	Fibre	Strength	Diameter	Fibre	Strength	Diameter
36	3.23	8.96	82	5.30	6.80	128	4.06	7.76
37	6.20	6.39	83	4.93	6.95	129	3.17	8.91
38	3.75	7.92	84	4.75	7.05	130	4.12	7.43
39	3.33	8.43	85	4.53	7.17	131	4.11	7.91
40	3.47	8.40	86	4.32	7.38	132	4.10	7.54
41	3.70	9.01	87	4.48	7.36	133	4.59	7.23
42	3.77	9.03	88	4.58	7.25	134	4.42	7.41
43	5.16	6.92	89	4.71	7.04	135	4.59	7.36
44	5.10	6.93	90	4.81	6.98	136	4.23	7.73
45	4.93	6.96	91	4.70	6.95	137	3.21	8.09
46	4.53	7.09	92	3.54	8.29			

Table 4.2 (continued). Strength and diameter measurements for 137 carbon fibres tension-tested in air

fit to a Weibull model is demonstrated and echoed by the probability plot of Figure 4.4, in which the fitted $F(t_{(i)})$ are obtained using maximum likelihood estimates for α and β in

$$F(t) = 1 - \exp[-(t/\alpha)^\beta].$$

For these data $\hat{\alpha} = 4.34$ and $\hat{\beta} = 6.09$.

It transpires that these data conform to the lognormal and gamma models quite well; however, it is more difficult to justify these models on physical grounds. It is always worth remembering that there are several aspects to effective statistical modelling.

4.8   Goodness of fit

Most of the general goodness-of-fit tests apply to grouped data or where data are compared to a completely specified distribution, that is, the parameters are not estimated from the data but given hypothesized values. They all focus on a measure of the difference between the empirical distribution of the sample and what is put forward as the theoretical distribution for the population. A point worth noting at the outset is that all goodness-of-fit tests are complicated in the presence of censoring.

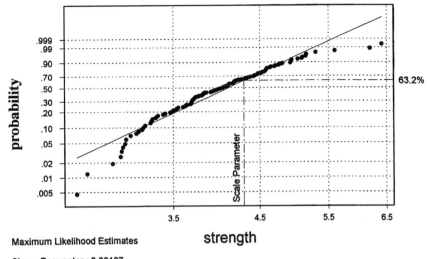

Maximum Likelihood Estimates

Shape Parameter : 6.09197
Scale Parameter : 4.33625

Figure 4.3. Weibull plot for Example 4.12

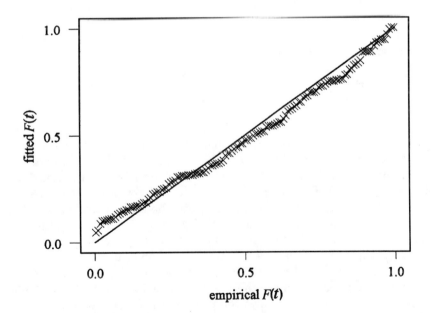

Figure 4.4. PP plot for Example 4.12

4.9   Pearson's χ^2 test

This is usually the first example of a goodness-of-fit test in most statistics courses. It requires a relatively large number of uncensored observations which are grouped into several observation categories. The latter may be quantitative or qualitative. The number of observations in each category is compared with the expected number of observations in each category under the assumed probability mechanism. For example, the number of breakdowns per shift in a process may be thought random and therefore constitute a Poisson process. Over a large number of shifts the proportion of shifts having x breakdowns would be compared with the expected proportion $e^{-\mu}\mu^x/x!$ (see Section 2.2), where μ is the mean number of breakdowns per shift. Now, the true value of μ is not likely to be known but would be estimated from the data. Due account is taken in the goodness-of-fit test of the uncertainty introduced by estimating unknown parameters.

Let there be k categories of observation. Observations may cover a numerical range, which is subdivided, not necessarily into equal intervals. Suppose there are o_i observations in category i. If under the assumed model there is a probability p_i that an observation falls in category i then the expected number of observations in category i, e_i, is np_i where n is the total number of observations. Provided all the e_i are at least 5 (and categories may be combined to achieve this) then the goodness-of-fit statistic

$$X^2 = \sum_{i=1}^{k} \frac{(o_i - e_i)^2}{e_i} \tag{4.8}$$

has approximately a χ^2 (*chi-squared*) distribution with a parameter known as the *degrees of freedom*, which here equals $k - 1 - r$, where r is the number of unknown parameters that have been estimated from the data. In the Poisson example above the value of r would be 1. Where the fit of the data to the model is not good the differences between the e_i and the o_i will be large, so a value of the test statistic exceeding the upper tail points of the χ^2 distribution indicates lack of fit.

The test will be illustrated using the carbon data in Table 4.2, but is not a recommended method for assessing the fit of models like the Weibull or lognormal where there are much better methods available. However, Example 4.14 shows the usefulness of the test where the p_i relate to discrete events and in particular to non-standard situations.

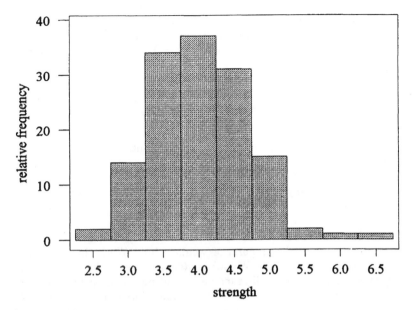

Figure 4.5. Histogram of the data in Table 4.2

Example 4.13: Figure 4.5 shows the data of Table 4.2 in the form of a *histogram*, where the area of the bars is proportional to the p_i, based on grouping the data as needed for the Pearson test. In a model fitting context, the histogram itself is of limited use. For these data, the normal, lognormal, gamma and Weibull families, among others, would be possibilities on the basis of this plot alone. To calculate the p_i under an assumed Weibull model we use

$$p_i = R(t_1; \hat{\alpha}, \hat{\beta}) - R(t_2; \hat{\alpha}, \hat{\beta})$$

$$= \exp\left\{-(t_1/\hat{\alpha})^{\hat{\beta}}\right\} - \exp\left\{-(t_2/\hat{\alpha})^{\hat{\beta}}\right\},$$

where in general t_1, t_2 are respectively the lower and upper ends of the *i*th grouping. For the first group t_1 is zero and for the last group t_2 is infinity. This ensures that the total expected number of observations matches the actual observed number. For the carbon data we have

interval	2.5	3.0	3.5	4.0	4.5	5.0	5.5	6.0	6.5
mid-point									
o_i	2	14	34	37	31	15	2	1	1
$e_i = np_i$	8.3	13.4	24.5	34.0	32.6	18.6		5.6	

The last two groups have very small e_i and in that case groups are combined in order to give all e_i with a value of at least 5. The number of groups is now 7 and since two parameters were estimated from the data, the number of degrees of freedom for the χ^2 test is $7 - 1 - 2 = 4$. The test statistic is

$$X^2 = \frac{(2-8.3)^2}{8.3} + \frac{(14-13.4)^2}{13.4} + \ldots + \frac{(4-5.6)^2}{5.6} = 9.99.$$

The upper 5% and 1% points on the χ^2 distribution with 4 degrees of freedom are 9.49 and 13.28. So the p-value lies between 0.01 and 0.05. This means that if the Weibull distribution adequately models the population, there is only a small probability that this set of observations would arise.

Example 4.14: The following is taken from Wolstenholme (1995) and concerns testing for the property of 'weakest link'. While this characteristic is associated with the Weibull distribution, it is not an exclusive relationship. The weakest link may exist but not necessarily mean that the random variable is Weibull. Twenty-four fibres, approximately 30 centimetres in length, were selected at random from the same fibrous material. Each was mounted in a frame which allowed sub-lengths of 5, 12, 30 and 75 millimetres to be separately tensile-loaded. The order in which the sub-lengths failed was observed. The weakest-link property would imply that the longer lengths generally fail first, because the more material there is the higher the number of potential failure sites. So, for example, the order of failure 75, 30, 12, 5 would be far more common than order 5, 12, 30, 75.

The different failure scenarios are grouped (fairly extensively because the number of fibres is just 24), into events E_1, E_2, E_3, and under the model described the expected proportion of observations in each group calculated. The results are as follows:

failure event	E_1	E_2	E_3
o_i	6	16	2
e_i	6.65	9.4	7.95

This yields

$$X^2 = \frac{(6-6.65)^2}{6.65} + \frac{(16-9.4)^2}{9.4} + \frac{(2-7.95)^2}{7.95} = 9.15.$$

The values of the p_i were calculated independently of the data, that is, the probability distribution was completely specified, and therefore $r = 0$ and the degrees of freedom $3 - 1 = 2$. The upper 5% point of χ^2 with 2 degrees of freedom is 5.99 and the upper 1% point 9.21. This implies that under the weakest-link assumption these data are fairly unusual, with probability (p-value) between 0.05 and 0.01. There is therefore some evidence against the model assumption and it is concluded that the weakest-link property does not apply to these data.

4.10   Kolmogorov–Smirnov test

There are a number of goodness-of-fit tests on the general lines of that due to Kolmogorov and Smirnov. Here the measure of the difference between the empirical distribution function and the proposed model is based on the maximum observed 'distance' between the two functions. The disadvantage of the basic Kolmogorov–Smirnov test is that the proposed model must be completely specified, in other words the parameters stated, not estimated. Estimating the model parameters introduces factors which are model-dependent, and specific examples of how this may be allowed for are shown in later sections.

Consider the goodness-of-fit problem where a sample of n uncensored observations is to be compared to a specified model based, say, on historic data. The Kolmogorov–Smirnov statistic is

$$D_n = \max | F_n(t_{(i)}) - F_0(t_{(i)}) | \tag{4.9}$$

where F_n is the sample distribution function and F_0 is the theoretical, completely specified distribution function, which may be discrete or continuous. The value of the test statistic will be large if the sample is very dissimilar to the proposed model. Critical values of D_n are based on the sample size and the level of significance. Examples are shown below for a 5% level of significance, and the hypothesized distribution is rejected if D_n exceeds the critical value:

n	5	10	20	50
critical value	0.5633	0.4092	0.2941	0.1884

Compared to the Pearson test, the basic D_n statistic has the disadvantage of requiring the parameters to be specified. However, because of the simplicity of D_n it is possible to reverse the process and, starting with the critical values of D_n, place confidence limits on the true distribution function, as shown, for example, in Cox and Lewis (1966).

Example 4.15: The following data come from an example in Chapter 5 of Leitch (1995) and concern lifetimes believed to come from an exponential distribution with rate parameter $\lambda = 0.01$:

$$3, 20, 40, 52, 53, 54, 85, 318, 429, 553$$

In keeping with Section 3.2, Leitch gives $F_n(t_{(i)}) = (i - 0.3)/(n + 0.4)$ due to the small sample size. The function $F_0(t)$ is given by $1 - \exp(-0.01t)$. Figure 4.6 shows these two distribution functions and the maximum deviation, which is 0.218. For sample size $n = 10$ the critical values are 0.40925 at a 5% level of significance and 0.48893 at 1%. The data are well within the region for accepting the hypothesized model.

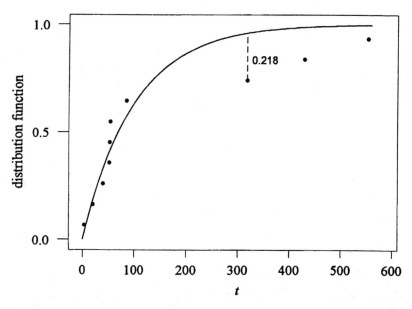

Figure 4.6. Cumulative distribution function plot for the Kolmogorov–Smirnov test in Example 4.15

A useful version of the Kolmogorov–Smirnov test applies to testing for differences of any kind between the populations from which two independent samples of uncensored observations have been drawn. Because the test is a non-parametric test of a very general hypothesis, the test is not as powerful (that is, sensitive to sample differences) as some other parametric, more specific tests. On the general topic of testing for differences between samples, the reader is referred to texts such as Chatfield (1983) and Hines and Montgomery (1990).

4.11 Tests for normality

Both Pearson's and the Kolmogorov–Smirnov test may be harnessed for the specific task of testing the fit of a normal distribution. Pearson allows estimation of population parameters but relies on a sufficient quantity of data. The Kolmogorov–Smirnov test does not in general allow for parameter estimation but there is a specific modification to allow for this in the normal distribution case. The population mean and standard deviation are estimated using (3.7) and (3.8). The data are then transformed using $z_{(i)} = (t_{(i)} - \bar{t})/s$ and the distribution function of the $z(i)$ compared to $N(0, 1)$. Again D_n is the maximum deviation between the two functions. Critical values of D_n are found in tables such as Neave (1985).

A long-standing and more favoured test for normality is that due to Shapiro and Wilk (1965). The test is based on an uncensored sample, and the test statistic is given by

$$W = \frac{\left(\sum_{i=1}^{n} a_i t_{(i)}\right)^2}{\sum_{i=1}^{n} (t_i - \bar{t})^2}, \tag{4.10}$$

where the a_i are functions of the sample size and distributional characteristics of the normal distribution. Values of these coefficients are tabulated, for example, in Pearson and Hartley (1972). In constrast to other tests, it is *small* values of W which indicate departure from normality.

Given now the availability of goodness-of-fit tests in software packages, the need to compute these statistics manually is reduced. MINITAB offers Kolmogorov–Smirnov, Ryan–Joiner (which is similar to Shapiro–Wilk) and Anderson–Darling (discussed in Section 4.12). Figure 4.7 shows a normal plot for logarithms of the strengths in Table 4.2. The plot seems to indicate a very good fit to these strengths by a

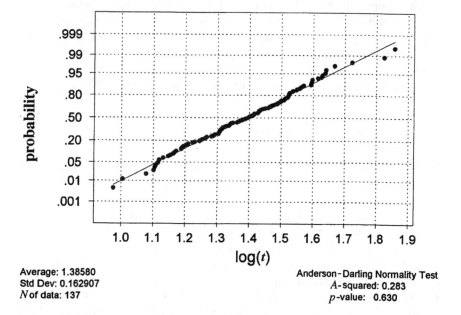

Average: 1.38580
Std Dev: 0.162907
N of data: 137

Anderson-Darling Normality Test
A-squared: 0.283
p-value: 0.630

Figure 4.7. Lognormal plot for the data in Table 4.2

lognormal model. The goodness-of-fit tests applied by MINITAB yield the following, and all give similar supporting evidence.

Test	statistic	p-value
Kolmogorov–Smirnov	0.038	>0.15
Ryan–Joiner/Shapiro–Wilk	0.9964	>0.1
Anderson–Darling	0.283	0.630

4.12 A^2 and W^2 tests

Two favoured tests based, like Kolmogorov–Smirnov, on the empirical distribution function are those due to Cramér and von Mises and to Anderson and Darling. With a completely specified F_0 these tests are also distribution-free, but in that case their use is again limited. However, much work has been done on the distribution of the test statistics when parameters are estimated for certain models. The computational form of the Cramér and von Mises statistic is given by

$$W^2 = \sum_i [F(t_{(i)};\hat{\theta}) - \hat{F}(t_{(i)})]^2 + \frac{1}{12n}. \tag{4.11}$$

This is principally a measure of the mean squared deviation of the distance between the data and the model with a correction based on the sample size, n. The Anderson–Darling statistic is computed as

$$A^2 = -2\sum_i \hat{F}(t_{(i)})[\log F(t_{(i)};\hat{\theta}) + \log R(t_{(n+1-i)};\hat{\theta})] - n. \tag{4.12}$$

In both (4.11) and (4.12) the Hazen formula is assumed for $\hat{F}$. Determination of $\hat{\theta}$ would usually be by maximum likelihood. A^2 is considered to have the advantage of taking account of the lack of variability in the tails of a distribution.

In work by Stephens (1977), the distributions of these statistics were investigated by Monte Carlo simulation and critical values determined for assessing the fit of either the Gumbel or Weibull models, the former being applicable to the logarithms of Weibull data. Where the two model parameters are both estimated, it is recommended that (4.11) and (4.12) are scaled by $(1 + 0.2/\sqrt{n})$. When the Weibull shape parameter is fixed, that is, the Gumbel scale parameter is fixed (of particular interest if testing the fit of an exponential model), then W^2 and A^2 are scaled by $(1 + 0.16/n)$ and $(1 + 0.3/n)$, respectively. Lack of fit is indicated by values exceeding those shown in Table 4.3.

	scaling	5% significance	1% significance
W^2	$1 + 0.2/\sqrt{n}$	0.124	0.175
	$1 + 0.16/n$	0.222	0.338
A^2	$1 + 0.2/\sqrt{n}$	0.757	1.038
	$1 + 0.3/n$	1.321	1.959

Table 4.3. Critical values for the W^2 and A^2 tests

Example 4.16: It was shown in Example 4.12 that the strength data of Table 4.2 did not conform to any plausible extent to a Weibull distribution. Applying the scaled forms of (4.11) and (4.12) yields $W^2 = 0.195$ and $A^2 = 1.628$. By reference to the critical values above, the poor fit of the Weibull model is confirmed. More complex approaches to modelling the lifetimes in this and similar data sets are discussed in Chapter 10.

4.13   Stabilized probability plots

Most plotting procedures suffer to some degree from points having different variabilities. In the probability plot described in Section 4.7 the points at the extremes have the lowest variability and a method for stabilizing the variability of points in this plot was first introduced by Michael (1983). Instead of plotting

$$(\hat{F}(t_{(i)}), F(t_{(i)}; \hat{\theta}))$$

where F is the proposed model and $\hat{\theta}$ the model parameters estimated usually by maximum likelihood, the distribution functions are transformed trigonometrically and the plotted points given by

$$\left(\frac{2}{\pi}\sin^{-1}\{\hat{F}(t_{(i)})\}^{1/2}, \frac{2}{\pi}\sin^{-1}\{F(t_{(i)}; \hat{\theta})\}^{1/2}\right).$$

Various authors have since constructed formal goodness-of-fit tests based on the stabilized probability plot. For example, Coles (1989) deals with the two-parameter Weibull distribution.

4.14   Censored data

Modification of goodness-of-fit tests to take account of censoring is simplest where all left censoring is at the lowest observation or all right censoring at the highest observation. Given that a range for the observed random variable is known, such as $[0, \infty]$, Pearson's χ^2 will adjust automatically. The statistics D_n, W^2 and A^2 are capable of modification but only remain independent of the model chosen for F_0 when the model is completely specified. The more general case of arbitrary censoring, that is, censored observations at various points during the interval of observation, is an unsolved problem, with or without parameter estimation.

Censoring combined with parameter estimation is best handled via techniques such as the likelihood ratio test, discussed in Chapter 9 and referred to in general terms in Section 4.4. A thorough discussion of goodness-of-fit may be found in Chapter 9 of Lawless (1982), and D'Agostino and Stephens (1986) provide a comprehensive reference on techniques appropriate for lifetime data.

CHAPTER 5

Repairable Systems

Every moment a man dies,
Every moment a man is born

— Alfred, Lord Tennyson (1809–1892)

5.1 Introduction

A repairable system is one which after failing to function properly may be restored to satisfactory working order by replacing or repairing certain components. The system will, throughout its lifetime, undergo alternating in-service times and repair/maintenance times. Components may be replaced at some point before failure, where excessive wear is likely to impair performance. Of interest are the times of failure and the cost incurred, both of which may be plotted against time to reveal patterns and any unusual features of the data. Repair/maintenance times may also be of interest, but it will be assumed initially that these are small relative to the in-service times.

If T_1, T_2, ... are the times of failure of the system, and X_i the time between the $(i - 1)$th and ith failures, we may take 'time' to be operating time, or real time if the 'down' times are considered to be negligble. What is of concern in observing the system over time is whether there is any change in the nature of the X_i . There will be no change if the X_i are independent and identically distributed. Departure from this condition may constitute *trend* and the detection and estimation of trends is a major interest. Even if there is no long-term trend there may be some other structure of practical importance, for example inadequate repairs may produce a run of short failure times among longer times. So there may be dependence or *autocorrelation* between the X_i.

5.2   Graphical methods

Three functions of prime interest relating to repairable systems are:

(i)   the mean cumulative repair function (MCRF);
(ii) the mean cumulative cost function (MCCF);
(iii) the rate of occurrence of failures (ROCOF).

Plots of the sample equivalents of these measures often reveal a simple underlying power law describing the evolution of the measures with time. If only one system is under study then the MCRF is simply the cumulative number of failures. If more than one system applies then the number of repairs is averaged over the systems in operation. The MCRF may have to be accumulated in stages if the number of systems in operation changes over time, say due to retirement.

The MCRF or MCCF may be plotted directly against time, or perhaps more usefully, the log MCRF or log MCCF against log t. Then an underlying power law may be estimated by drawing a straight line through the points.

Let $N(t)$ be the number of failures in the time interval $(0, t)$. Then

$$\text{MCRF} = E[N(t)].$$

Writing this in power law form as

$$E[N(t)] = At^{\alpha}$$

we have

$$\log \text{MCRF} = \log A + \alpha \log t.$$

A plot of log MCRF against log t, known as a Duane plot (Duane 1964), first reveals whether a power law is a suitable model for the MCRF and second provides an estimate of the parameter α via the slope of the fitted line. Signs of non-linearity may indicate an inadequate model, or features such as a 'kink' in the line may indicate a change in the system due, for example, to an effective quality improvement programme.

If $C(t)$ is the cumulative cost at time t, then similarly

$$\text{MCCF} = E[C(t)] = Bt^{\beta},$$

and

$$\log \text{MCCF} = \log B + \beta \log t.$$

The ROCOF, or *intensity function*, is defined by

$$\lambda(t) = \frac{d}{dt}\{E[N(t)]\}$$

$$= A\alpha t^{\alpha-1}.$$

Thus if $0 < \alpha < 1$, the ROCOF decreases; if $\alpha = 1$, the ROCOF is constant; and if $\alpha > 1$, the ROCOF increases. Ascher and Feingold (1984) coined the terms 'happy system' and 'sad system' where the ROCOF is decreasing and increasing, respectively.

The existence of trend is therefore characterized by linear plots which have slope significantly different from 1. An approximate measure of the possible departure from 1 can be gained via a confidence interval estimate of the slope as described in the regression analysis for Weibull plots in Section 4.3.

The ROCOF should not be confused with the hazard function. The hazard describes the conditional probability of failure at time t of a non-repairable lifetime. It is possible for each X_i to have an increasing hazard but for the system to have a constant ROCOF, or indeed any other combination of behaviours. A battery may have Weibull distributed lifetime with shape parameter greater than 2, implying increasing hazard, but a system in which the battery is replaced on failure by a similarly performing battery will display constant ROCOF.

Example 5.1: The following data (attributed to A. Winterbottom) concern repairs to a large number (945) of motor vehicles in a warranty period. The values shown are months in service (MIS) and cumulative repairs per unit and cumulative cost per unit, that is the measures have been averaged over the 945 vehicles.

MIS	1	2	3	4	5	6	7	8
R (MCRF)	0.397	0.547	0.688	0.807	0.950	1.064	1.173	1.305
C (MCCF)	11.84	16.44	20.84	24.68	29.73	32.76	36.82	41.14

Figure 5.1 shows $\log R$ and $\log C$ plotted against $\log$ MIS. Both plots approximate to a straight line with *coincidentally* similar slopes. The power estimated by the slope is approximately 0.6, thus both repairs and costs are showing a decreasing rate of occurrence over the period.

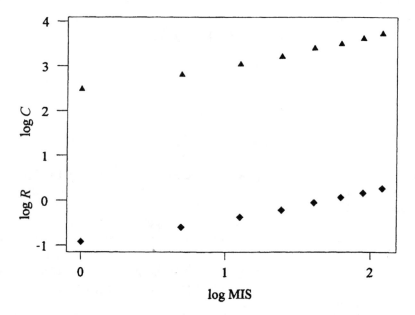

Figure 5.1. MCRF and MCCF plots for Example 5.1

Example 5.2: This example is from an article by Nelson (1988) concerning repair and cost data for six repairable systems, all commissioned at time zero but each subject to retirement over time. The data cover a period of 42 months. Figure 5.2 illustrates the data, each event shown as a repair (×) or retirement (o) with the cost of repair (in hundreds of dollars). Calculation of the MCRF and MCCF is shown in Table 5.1.

Figures 5.3 and 5.4 show log time plotted against log MCRF and log MCCF. Both plots are linear, indicating that a power law model is appropriate for both MCRF and MCCF. The slopes of the graphs indicate how the rate of repair and costs are varying over time. In the case of MCRF the gradient is approximately one and therefore the repair rate is fairly constant. The costs show a rate of change of 1.16 which indicates increasing costs, but given the length of time over which the data were collected, this factor may be purely due to inflation. This latter point shows that it is particularly important when costs are considered to measure them in the most appropriate way. Over long periods of time it may be necessary to index them or perhaps measure them in terms of different units, such as person-hours.

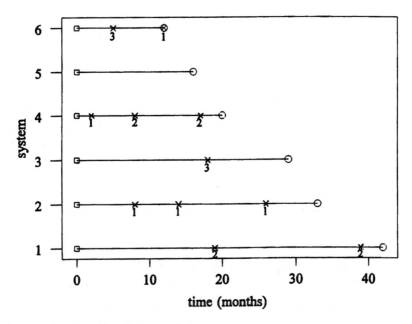

Figure 5.2. Repair data for Example 5.2

Event	Time	Cost	Systems in service	Mean repairs	MCRF	Mean cost	MCCF
1	2	1	6	1/6	0.17	1/6	0.17
2	5	3	6	1/6	0.33	3/6	0.67
3	8	1	6	1/6	0.50	1/6	0.83
4	8	2	6	1/6	0.67	2/6	1.17
5	12	1	6	1/6	0.83	1/6	1.33
6	12	–					
7	14	1	5	1/5	1.03	1/5	1.53
8	16	–					
9	17	2	4	1/4	1.28	2/4	2.03
10	18	3	4	1/4	1.53	3/4	2.78
11	19	2	4	1/4	1.78	2/4	3.28
12	20	–					
13	26	1	3	1/3	2.11	1/3	3.61
14	29	–					
15	33	–					
16	39	2	1	1/1	3.11	2/1	5.61
17	42	–					

Table 5.1. Calculation of MCRF and MCCF for Example 5.2

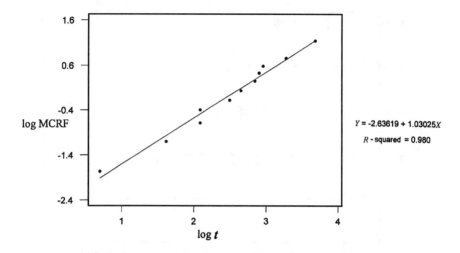

Figure 5.3. Duane plot for Example 5.2 repair data

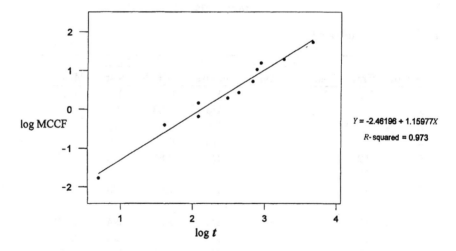

Figure 5.4. Duane plot for Example 5.2 cost data

5.3   Testing for trend

A process in which the ROCOF varies with time may be described as a *non-homogeneous Poisson process*. The function $\lambda(t)$ may be a power law, say $A\alpha t^{\alpha-1}$, or perhaps log linear, expressed as $\exp(\alpha+\beta t) = Be^{\beta t}$

testing H_0: no trend, against H_1: decreasing ROCOF, the critical n at the 2.5% significance level is $u < -1.96$, so H_0 is rejected at this and it is concluded that the incidence of failure is tending to reduce.

mple 5.4: Proschan (1963) gives the following 13 observed intervals erating hours between failure of the air-conditioning equipment in eing 720 aircraft:

413, 14, 58, 37, 100, 65, 9, 169, 447, 184, 36, 201, 118.

cumulative times of failure are given by

3, 427, 485, 522, 622, 687, 696, 865, 1312, 1496, 1532, 1733, 1851.

refore $T_n = 1851$, where $n = 13$. Accumulating all T_i up to $i = 12$ gives tal of 10790 so

$$u = \frac{10790/12 - 1851/2}{1851\sqrt{\frac{1}{12 \times 12}}} = 0.4776.$$

ure 5.5 shows the log MCRF plotted against log t. The linearity of plot is not good and would not be improved by using t instead of log there is no straightforward model for the repair function. The modest ue of u shows that there is no significant overall trend in the occurance of repairs. There is some suggestion of cycles in the data, but more servations would be needed to identify any firm pattern.

ample 5.5: Continuing Example 5.4, the next 10 failures were served after the equipment had a major overhaul. The intervals were

34, 31, 18, 18, 67, 57, 62, 7, 22, 34.

gure 5.6 shows a plot of cumulative number of failures against cumutive time. This again hints at some cyclic behaviour but also clearly ows that the effect of the overhaul appears to be an increase in the te of occurrence of failure. A formal trend test for the complete data ould confirm this. It may, however, be of interest to know what the rocess behaviour was after overhaul. The ROCOF may have increased, ut has it moved to a new constant level?

the time origin is now reset to zero at the 13th overhaul, the cumulative mes for the next 10 failures are

34, 65, 83, 101, 168, 225, 287, 294, 316, 350.

which can also model both happy ($\beta < 0$) and
number of failures in (t_1, t_2) has a Poisson di

$$\mu_{(t_1, t_2)} = \int_{t_1}^{t_2} \lambda(t)\mathrm{d}t,$$

and the probability of no failures in (t_1, t_2) is

$$e^{-\mu_{(t_1, t_2)}} = \exp\left\{-\int_{t_1}^{t_2} \lambda(t)\mathrm{d}t\right\}.$$

The Laplace test for trend

Under the assumption of constant λ, the time
T_{n-1} are ordered uniform variates (see Example
that this is conditional on the observed value
distribution on [0, a] has mean $a/2$ and varianc
By the central limit theorem (see Sectio
$\sum_{i=1}^{n-1} T_i/(n-1)$ has an approximate normal dis
$T_n/2$ and variance $T_n^2/[12(n-1)]$. So

$$U = \frac{\displaystyle\sum_{i=1}^{n-1}\frac{T_i}{n-1} - \frac{T_n}{2}}{T_n\sqrt{\dfrac{1}{12(n-1)}}}$$

is approximately $N(0, 1)$, and may be used to test
(H_0) of no trend, that is, $\beta = 0$ in $\beta \exp(\beta t)$ or $\alpha =$
increasing ROCOF ($\beta > 0$ or $\alpha > 1$), decreasing RO(
or either ($\beta \neq 0$ or $\alpha \neq 1$).

Example 5.3: Failures occur at times 15, 42, 74, 117
(5.1), the value of u is

$$\frac{(15 + 42 + 74 + 117 + 168 + 233)/6 - 410/2}{410\sqrt{\dfrac{1}{12 \times 6}}} =$$

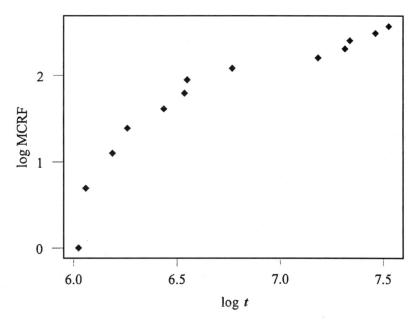

Figure 5.5. Duane plot for Example 5.4

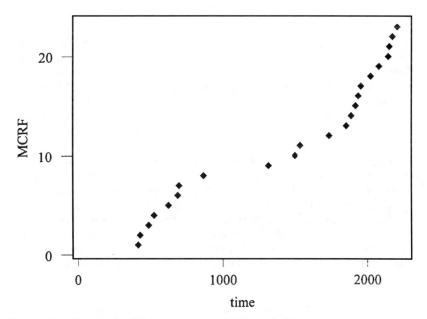

Figure 5.6. Cumulative failures over time for Example 5.5

The value of T_n is 350 and $n = 10$. So

$$u = \frac{1573/9 - 350/2}{350\sqrt{1/12 \times 9}} = -0.0066$$

This very small value of u implies no evidence of trend. The major overhaul has clearly changed the ROCOF but the change is from one constant level to another, higher, constant level. Further investigation into the precise cause of failure may provide a reason, for example some imperfect repair at overhaul or use of a low-quality component.

In software reliability the elimination of faults is of interest. Over time a growth in reliability is required, that is, evidence that the ROCOF is decreasing. The question of how many faults remain at any time is an important question which will not be addressed here.

Example 5.6: Here we consider data first analysed by Jelinski and Moranda (1972). The following data show the intervals in days between successive failures of a piece of software.

9	12	11	4	7	2	5	8	5	7	1	6
1	9	4	1	3	3	6	1	11	33	7	91
2	1	87	47	12	9	135	258	16	35		

Figure 5.7 shows a plot of log MCRF against log t. A power law model for trend seems appropriate, but a change in its nature around the 21st failure is evident. Up to this point a power law model with $\alpha = 1.3$ fits very well, and indicates a slightly increasing ROCOF. If after 21 failures we reset the time origin as before, we find that a power law model is again appropriate, but now with $\alpha = 0.74$. There is now a significantly decreasing rate of occurrence of failure.

Example 5.7: In order to adapt the Laplace test to a situation where multiple systems are in operation and may be subject to retirement, or where systems are commissioned at different times, the time scale used should be one reflecting total unit operating time. Consider Example 5.2 again. The total operating time until the first repair is 6×2 months. There are a further 6×3 operating months until the next repair, and so on. Between events 7 and 9 there are $5 \times 2 + 4 \times 1$ operating months taking into account the retirement of system 5. The data to be applied to a Laplace trend test are thus

12, 30, 48, 48, 72, 82, 96, 100, 104, 126, 149.

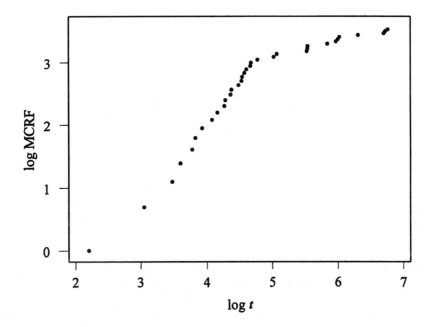

Figure 5.7. Duane plot for Example 5.6

Strictly speaking, the Laplace test is based on the null hypothesis that the process is a homogeneous Poisson process. However, the test may be used for the more general hypothesis that the process is a *renewal process* (see Section 5.6), that is, the inter-event times are IID. Lews and Robinson (1974) recommend that in this case the U statistic be divided by the *coefficient of variation* of the inter-event times, which is estimated by the sample standard deviation divided by the sample mean. The exponential distribution has a coefficient of variation equal to 1, so this modification embraces the U of (5.1).

The formulation of the Laplace test rests on the central limit theorem, which would imply that the test is only good when the value of n is large. However, the test proves to be fairly robust against departure from normality (Bates 1955) and may be applied even when n is modest. Ascher and Feingold (1984) give a full account of analytical trend tests.

There are a number of models for the rate of occurrence of failure and hence for reliability 'growth'. In the power law relationship $MCRF = At^{\alpha}$, a value of α significantly less than one will indicate reliability growth. The reader is referred to Chapter 7 of Ansell and Phillips (1994) for a comprehensive account of growth models.

5.4   Repair time

One way of describing the occurrence of failures is to state the *mean time between successive failures*, often known as the MTBF. The main difference between the MTBF and the *mean time to failure*, known as the MTTF, is in the usage. The MTTF may apply to first failure or a non-repairable failure, whereas the MTBF is used in the context of repairable systems. Note that the MTBF is usually the mean *operating* time between failures. There is, however, considerable variation in the use of these terms and the definitions should be checked within any given text.

Repair actions taken *after* system or component failure are *corrective maintenance* actions. Actions taken *before* failure, or *preventive mainte- nance*, may include regular servicing and replacement of parts before significant wear or total failure has occurred.

The modelling of repair time is of considerable interest as this affects the necessary human resources, spares provisioning and oper- ational effectiveness. Modelling of maintenance activity is considered in Chapter 8.

5.5   Maintainability and availability

Maintainability is defined to be the probability that a component or system that has failed will be restored to full working order within a specified period of time τ.

Times are frequently assumed to be exponentially distributed, and it is only in this case that rate of failure or rate of repair is given by the reciprocal of mean time to failure or mean time to repair (MTTR). In this case, the maintainability $M(\tau)$ may be expressed as

$$M(\tau) = 1 - \exp(-\tau/\text{MTTR}).$$

Reliability and maintainability are combined in the term *availability* which is the proportion of 'up' time. The long-term or 'steady-state' availability, A_∞, is given by

$$A_\infty = \frac{\text{MTBF}}{\text{MTBF} + \text{MTTR}}.$$

It should be noted that A_∞ provides no information about the number of repairs.

Frequently 'down' time involves more than the time to repair. Factors might include time to diagnose repair, parts location and testing, for example. Expressions for availability therefore often involve mean down time MDT rather than MTTR. Given exponential lifetimes with hazard λ, the *unavailability* is given by

$$\bar{A}_\infty = 1 - A_\infty = \frac{\lambda \text{MDT}}{1 + \lambda \text{MDT}}.$$

For low hazard rate and small MDT the unavailability is approximately λMDT.

5.6 Introduction to renewal theory

Let it be assumed that the failure time of components is a random variable with probability density function $f(x)$. Suppose that a new component is put into operation at time zero. When this component fails at time X_1, it is replaced immediately by a new component whose time to failure is X_2, and so on. If X_1, X_2, ... are IID random variables we call the system an *ordinary renewal process* (ORP).

Suppose, however, that the component in use at time zero is not new, that is the time origin is after the start of the ORP. Then X_1 has a different probability density function (p.d.f.) from the other X_i. (Figure 5.8). This is called a *modified renewal process* (MRP).

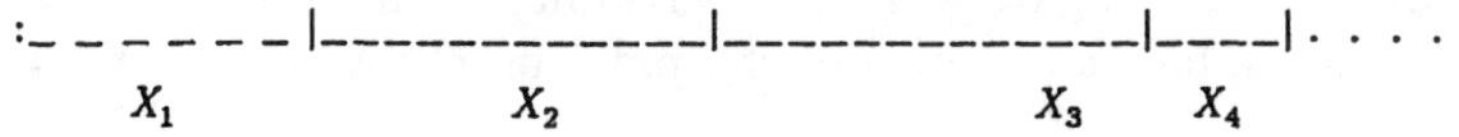

Figure 5.8. A modified renewal process

A number of variables associated with these processes are of interest. The time at which the rth renewal occurs is expressed as

$$S_r = X_1 + X_2 + X_3 + \ldots + X_r.$$

It may be possible to calculate the distribution of S_r exactly or we can use the central limit theorem. The number of renewals in $(0,\ t)$ is denoted by $N(t)$. This is related to S_r via

$$P(N(t) < r) = P(S_r > t).$$

The expected value of $N(t)$ is known as the *renewal function*:

$$M(t) = E[N(t)].$$

Differentiating with respect to time, we obtain the *renewal density*, or *first-order intensity*,

$$m(t) = \frac{\mathrm{d}}{\mathrm{d}t} M(t).$$

5.7   Laplace transforms

Let $f(x)$ be a function defined on $(0,\infty)$; then the *Laplace transform*, $L\{\,f(x)\,\}$, generally denoted $f^*(s)$, is defined (recalling (1.11)) by

$$f^*(s) = E(e^{-sX}) = \int_0^\infty e^{-sx} f(x)\mathrm{d}x. \tag{5.2}$$

In a statistical context, when $f(x)$ is a p.d.f. the Laplace transform is a *moment generating function* which, as its name implies, can be used to generate moments of the distribution.

Laplace transforms are applied in renewal theory in connection with sums of independent random variables. Let $X_1,\ X_2,\ ...,\ X_n$ be independent non-negative random variables with p.d.f.s $f_1(x),\ f_2(x),\ ...,\ f_n(x)$. The Laplace transform of the p.d.f. of $Y = X_1 + ... + X_n$ is

$$E[e^{-sY}] = E[e^{-sX_1} e^{-sX_2} ... e^{-sX_n}]$$

$$= E[e^{-sX_1}]E[e^{-sX_2}]...E[e^{-sX_n}] \text{ since the } X_i \text{ are independent,}$$

$$= f_1^*(s)f_2^*(s)...f_n^*(s). \tag{5.3}$$

A special case is when the f_i are identical, for then

$$f_Y^*(s) = [f_X^*(s)]^n,$$

and the p.d.f. of Y, $f_Y(y)$, is called the *n-fold convolution* of $f(x)$.
The Laplace transform is unique:

$$f_1^*(s) = f_2^*(s) \Leftrightarrow f_1^*(x) = f_2^*(x).$$

The problem of finding $f(x)$ from $f^*(s)$ is called the *inversion problem*.
Inversion formulae are available, but standard results such as those
in Table 5.2, combined with partial fractions, enable many cases to be
solved. It may also be possible to expand a Laplace transform as a
power series in s using, say, Taylor expansions, and then the function
may be inverted term by term.

$f^*(s)$	$f(x)$
$(s + a)^{-n}$	$e^{-ax} x^{n-1}/(n-1)!$
$[(s + a)(s + b)]^{-1}$	$(e^{-ax} - e^{-bx}) / (b - a)$

Table 5.2. Laplace transforms and corresponding functions

For the purposes of this text it is also useful to have the following
result:

$$L\left\{\frac{d}{dt} f'(t)\right\} = s f^*(s) - f(0). \tag{5.4}$$

If f is discontinuous at $t = 0$ then $f(0)$ is taken to be the limit of $f(t)$
as t tends to zero from positive t.

5.8   The renewal function

Let $K_r(t)$ be the cumulative distribution function of the time to the rth
renewal, S_r. Recall that

$$P(N(t) < r) = P(S_r > t);$$

so

$$P(N(t) < r) = 1 = K_r(t).$$

Then

$$P(N(t) = r) = K_r(t) - K_{r+1}(t),$$

where r is an integer. Now

$$M(t) = E[N(t)] = \sum_{r=0}^{\infty} rP(N(t) = r)$$

$$= \sum_{r=0}^{\infty} r[K_r(t) - K_{r+1}(t)]$$

$$= \sum_{r=1}^{\infty} K_r(t)$$

Differentiating with respect to t,

$$m(t) = \sum_{r=1}^{\infty} k_r(t),$$

where $k_r(t)$ is the p.d.f of S_r. It then follows from (5.3) that

$$m^*(s) = \sum_{r=1}^{\infty} f_1^*(s)f_2^*(s)\ldots f_r^*(s).$$

Thus for an ORP,

$$m^*(s) = \sum_{r=1}^{\infty} [f(s)]^r = \frac{f^*(s)}{1 - f^*(s)}, \tag{5.5}$$

using the result for the sum of a geometric progression (see the Appendix) and assuming that $[f(s)]^r$ tends to zero. For an MRP,

$$m^*(s) = f_1^*(s) \sum_{r=1}^{\infty} [f(s)]^{r-1} = \frac{f_1^*(s)}{1 - f^*(s)}. \tag{5.6}$$

Example 5.8: An ORP for which $f(x) = \lambda e^{-\lambda t}$ has $f^*(s) = \lambda/(\lambda + s)$. From (5.5),

$$m^*(s) = \frac{\lambda}{\lambda + s} \Big/ \left(1 - \frac{\lambda}{\lambda + s}\right) = \lambda/s.$$

Inverting gives

$$m(t) = \lambda = \frac{1}{\text{mean}}.$$

From (5.4),

$$m^*(s) = sM^*(s) - M(0),$$

and since $M(0) = 0$, $M(t)$ is given by

$$M^*(s) = \frac{m^*(s)}{s}. \tag{5.7}$$

A distribution of particular interest in renewal theory is the *special Erlangian distribution*, which has p.d.f.

$$f(x) = \frac{\lambda^a x^{a-1} e^{-\lambda x}}{(a-1)!},$$

where a is an integer. It is a special case of the gamma distribution. It has particular application where failure takes place in a 'stages', the times spent in these stages being independently exponentially distributed with p.d.f. $\lambda e^{-\lambda y}$. The time to failure, $X = Y_1 + Y_2 + \dots + Y_a$, then has a special Erlangian distribution. A generalization is one in which the exponential distributions in the a stages have parameters $\lambda_1, \lambda_2, \dots, \lambda_a$.

For an ORP with failure times special Erlangian distributed $N(t) = r$ if and only if k, the number of stages completed in the underlying process, is such that $ra \le k < ra + (a - 1)$.

Given the Laplace transform of the exponential distribution, $f_X^*(s)$ for the generalized Erlangian distribution is

$$\frac{\lambda_1}{\lambda_1 + s} \frac{\lambda_2}{\lambda_2 + s} \frac{\lambda_3}{\lambda_3 + s} \cdots \frac{\lambda_a}{\lambda_a + s}.$$

5.9   Alternating renewal processes

Suppose there are two types of component with respective failure times $\{X_1, X_2, \dots\}$ and $\{Y_1, Y_2, \dots\}$, and all failure times are statistically independent. A process whereby, on failure, a component is replaced by one of the other type is called an *alternating renewal process* (Figure 5.9). An example of such processes is a two-unit system, whereby one unit is set in operation at time zero and when it fails it is replaced by the spare, or *standby*, unit and system failure is defined to be when the standby unit fails. Another is a machine subject to stoppage while being repaired, so there is an alternating sequence of 'in-service' and 'down' times, representing two sequences of independent random variables.

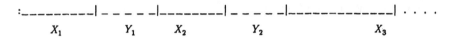

$$\;\;X_1 \qquad\qquad Y_1 \qquad X_2 \qquad\qquad Y_2 \qquad\qquad\qquad X_3$$

Figure 5.9. An alternating renewal process

Consider the ORP $\{Z_i\}$, where $Z_i = X_i + Y_i$. We have

$$f_Z^*(s) = f_X^*(s) f_Y^*(s).$$

The renewal function, that is, the expected number of failures in $(0, t)$, may be obtained from (5.5) and (5.7) by replacing $f^*(s)$ by $f_Z^*(s)$, giving

$$M^*(s) = \frac{f_X^*(s)\ f_Y^*(s)}{s\{1 - f_X^*(s)\ f_Y^*(s)\}}.$$

In the machine repair example above the failures are recorded at the end of the 'X' failure times, so the system has a sequence of failure times $Z_1, Z_2, \dots = X_1, Y_1 + X_2, Y_1 + X_2, \dots$, that is, an MRP. The renewal function is then given by

$$M^*(s) = \frac{f_X^*(s)}{s\{1 - f_X^*(s)\ f_Y^*(s)\}}. \tag{5.8}$$

Example 5.9: When X is exponential with parameter λ_1, and Y is exponential with parameter λ_2, (5.6) yields

$$m^*(s) = \frac{\lambda_1(\lambda_2 + s)}{\{s^2 + (\lambda_1 + \lambda_2)s\}}$$

$$= \frac{\lambda_1\lambda_2}{s(s + \lambda_1 + \lambda_2)} + \frac{\lambda_1}{s + \lambda_1 + \lambda_2}.$$

Inverting, using Table 5.2,

$$m(t) = \lambda_1\lambda_2\frac{(1 - e^{-(\lambda_1 + \lambda_2)t})}{(\lambda_1 + \lambda_2)} + \lambda_1 e^{-(\lambda_1 + \lambda_2)t}$$

$$= \frac{\lambda_1}{(\lambda_1 + \lambda_2)}(\lambda_2 + \lambda_1 e^{-(\lambda_1 + \lambda_2)t}).$$

At $t = 0$ the renewal density is λ_1, the renewal density of X_1, and as t tends to infinity $m(t)$ approaches $(\lambda_1\lambda_2)/(\lambda_1 + \lambda_2)$, which is the reciprocal of $1/\lambda_1 + 1/\lambda_2$, the mean of $X + Y$. Figure 5.10 illustrates $m(t)$ for $\lambda_1 = 1$ and $\lambda_2 = 2$.

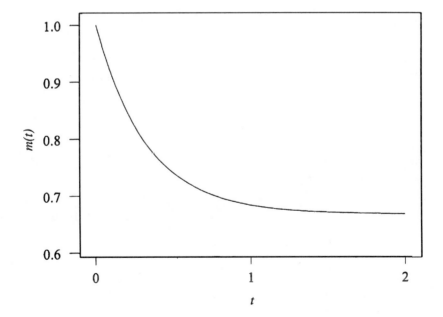

Figure 5.10. Renewal density for Example 5.9

One consequence of this analysis is that if spares provisioning is based on long-term behaviour, shortages are likely to occur in the short term.

5.10   The distribution of $N(t)$

It can be shown that $N(t)$ is asymptotically (that is, as $t \rightarrow \infty$) normally distributed with mean t/μ and variance $\sigma^2 t/\mu^3$, where μ and σ^2 are the mean and variance of the distribution of failure times. Various techniques are available which give a guide as to how large t must be before the approximation is adequate, but a 'steady-state' behaviour will not be achieved until a large number of renewals have taken place.

For further discussion of renewal theory the reader is referred to the classic text by Cox (1962).

CHAPTER 6

System Reliability

6.1 Systems and system logic

We can think of a *system* as an entity with components, parts or blocks, with lines of communication between them. An electric circuit, a piece of software, an organization are all examples of systems. A system can often be best represented by a diagram showing the constituent parts and their interconnections, often referred to as a *reliability block diagram* (RBD). Such a system can usually be described as having an input and an output, and a route through the system connecting the input to the output may be called a *path*. A system works if there is a successful path between the input and output. A path is successful if every component along the path does not fail.

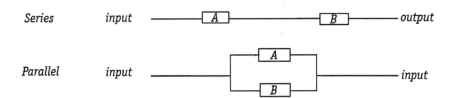

Figure 6.1. Reliability block diagrams for series and parallel systems of two components

There are two basic types of system structure, *series* and *parallel*. A series system works only if all components work. A parallel system works provided at least one of the components works. Figure 6.1 shows RBDs for simple series and parallel systems.

We can use the Boolean notation of probability theory to write

$$P(\text{series system works}) = P(A \text{ works} \cap B \text{ works}),$$

$$P(\text{parallel system works}) = P(A \text{ works} \cup B \text{ works}).$$

Boolean variables can be used to describe when a component is *up* (working) or *down* (not working). If the components are labelled $A_1, A_2, \ldots$

$$y_i = \begin{cases} 1 \text{ if } A_i \text{ is up,} \\ 0 \text{ if } A_i \text{ is down.} \end{cases}$$

A *structure function*, ϕ, can be used to describe whether the system is up or down as follows: for series, systems,

$$\phi = y_1 \, y_2 \, \ldots$$

while for parallel,

$$\phi = 1 - \{(1 - y_1)(1 - y_2) \ldots \}.$$

When $\phi = 1$ the system is working, where $\phi = 0$ the system is not working.

A system is described as *coherent* if it is one for which

(i)   each component is relevant, that is, has some effect on the system, and

(ii) ϕ is an increasing function of the y_i, that is, a component moving from not working to working does not have a negative effect on the system.

It is often possible to decompose a complex system, S, into different parts, or subsystems, A_i, written as

$$S = \bigcup_{i=1}^{n} A_i,$$

where the subsystems A_i are disjoint, that is, non-overlapping. Then if the reliability functions of the subsystems are evaluated, and the subsystems are combined say in series or parallel, then ϕ for the whole system can be found, as if the subsystems were separate components. It should be noted that some systems may not necessarily be simply described just in terms of series and parallel structures.

Another kind of decomposition is possible into groups of components that can fail or function together, in the sense that their failure or functioning can directly affect the functioning of the system. The most important such groups are *ties* and *cuts*, which are generalizations of the concepts of series and parallel systems.

6.2   Tie and cut sets

A tie set, or path, is a set of components joining the input to the output.
A branch joins two components. Of particular interest are the *minimal tie
sets* which are paths consisting of the smallest number of branches. If there
are n of these minimal tie sets $\{T_1, T_2, ..., T_n\}$ then the system reliability is

$$R = P(T_1 \cup T_2 \cup ... \cup T_n),$$

where T_i indicates that the tie is working, that is, all components on
the path are working.

A *minimal cut set* is similarly defined. It contains a minimal number of components which if removed from the system would cause the
system to fail. If there are m minimal cut sets $\{C_1, C_2, ..., C_m\}$ then
the system reliability is

$$R = 1 - P(C_1 \cup C_2 \cup ... \cup C_m),$$

where C_i indicates that the cut C_i has occurred, that is, the components
in the cut have all failed.

The analysis depends critically on whether we assume independence of failure. It may be that failure of individual components is
independent, but ties (paths) T_i or cuts C_i are not usually independent.

Example 6.1: Figure 6.2 shows a system with components A, B, C, D
which fail independently. Paths through the system are given by $\{ABD\}$
and $\{ACD\}$, both of which are minimal tie sets because there is no subset
of either path which is also a path. The cuts are $\{A\}$, $\{BC\}$ and $\{D\}$, again
minimal. The paths will respectively be denoted T_1, T_2 and the cuts C_1,
C_2, C_3. The system reliability is given by

$$
\begin{aligned}
R &= P(\text{system works}) \\
&= P(\text{at least one successful path}) \\
&= 1 - P(\text{at least one cut failed}).
\end{aligned}
$$

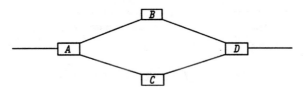

Figure 6.2. Block diagram for Example 6.1

So here,

$$R = P(T_1 \cup T_2) = 1 - P(C_1 \cup C_2 \cup C_3).$$

6.3   Probability bounds

Rule (1.3) can be extended to cover the union of any number of sets. Here we are interested in

$$R = P(T_1 \cup T_2 ... \cup T_n)$$
$$= \sum_i P(T_i) - \sum_{i<j} P(T_i \cap T_j) + ...(-1)^{n-1} P(T_1 \cap T_2 \cap ... \cap T_n). \quad (6.1)$$

Alternatively, it may be easier to calculate the unreliability

$$\bar{R} = P(\text{not working}) = P(C_1 \cup ... \cup C_m)$$
$$= \sum_i P(C_i) - \sum_{i<j} P(C_i \cap C_j) + ... + (-1)^{m-1} P(C_1 \cap C_2 \cap ... \cap C_m). \quad (6.2)$$

To put workable bounds on these probabilities we can use just the first few terms in these expansions. For example,

$$\sum_i P(T_i) - \sum_{i<j} P(T_i \cap T_j) \le R \le \sum_i P(T_i),$$
$$\sum_i P(C_i) - \sum_{i<j} P(C_i \cap C_j) \le \bar{R} \le \sum_i P(C_i).$$

If failure probabilities $\{p_j\}$ for components are small and independent, then we may take

$$\bar{R} \cong \sum_i \prod_{j \in c_i} p_j.$$

Example 6.1 (continued): It is easiest to use ties here because there are fewer ties than cuts, but we will demonstrate that the same result is obtained with either. We have

$$P(T_1 \cup T_2) = P(T_1) + P(T_2) - P(T_1 \cap T_2)$$
$$= P(ABC) + P(ACD) - P(ABCD).$$

Note that nodes A and D each occur only once in the last term. The ties are not independent and hence a straight product of the tie probabilities is not valid, but nodes are independent so, for example, $P(ABD) = P(A)\,P(B)\,P(D)$, where $P(A)$ is the probability that A works.

Suppose that A and D each fail with probability α and B and C each fail with probability β. Then

$$
\begin{aligned}
R &= (1-\alpha)(1-\beta)(1-\alpha) + (1-\alpha)(1-\beta)(1-\alpha) - (1-\alpha)(1-\beta)(1-\beta)(1-\alpha) \\
&= (1-\alpha)^2(1-\beta)[2-(1-\beta)] \\
&= (1-\alpha)^2(1-\beta)[1+\beta] \\
&= (1-\alpha)^2(1-\beta^2).
\end{aligned}
$$

Using cuts,

$$
\begin{aligned}
\bar{R} = {} & P(C_1) + P(C_2) + P(C_3) - P(C_1 \cap C_2) - P(C_1 \cap C_3) \\
& - P(C_3 \cap C_2) + P(C_1 \cap C_2 \cap C_3).
\end{aligned} \tag{6.3}
$$

Now expressing these cuts in terms of the failed components, we have

$$
\begin{aligned}
\bar{R} &= P(\bar{A}) + P(\bar{B}\bar{C}) + P(\bar{D}) - P(\bar{A}\bar{B}\bar{C}) - P(\bar{A}\bar{D}) - P(\bar{B}\bar{C}\bar{D}) + P(\bar{A}\bar{B}\bar{C}\bar{D}) \\
&= \alpha + \beta^2 + \alpha - \alpha\beta^2 - \alpha^2 - \beta^2\alpha + \alpha^2\beta^2 \\
&= 2\alpha - \alpha^2 + \beta^2[1 - \alpha - \alpha + \alpha^2].
\end{aligned}
$$

So,

$$
\begin{aligned}
R &= 1 - 2\alpha + \alpha^2 - \beta^2[1 - 2\alpha + \alpha^2] \\
&= (1-\alpha)^2(1-\beta^2), \text{ as above.}
\end{aligned}
$$

Example 6.2: Figure 6.3 is an RBD for part of a computer system. There are six ties for this system but just three cuts, $\{ABC\}$, $\{D\}$, $\{EF\}$, so clearly it is better to calculate reliability via cuts.

From (6.3) the unreliability is given by

$$
\begin{aligned}
\bar{R} &= P(\bar{A}\bar{B}\bar{C}) + P(\bar{D}) + P(\bar{E}\bar{F}) - P(\bar{A}\bar{B}\bar{C}\bar{D}) \\
&= P(\bar{D}\bar{E}\bar{F}) - P(\bar{A}\bar{B}\bar{C}\bar{D}\bar{E}\bar{F}) + P(\bar{A}\bar{B}\bar{C}\bar{D}\bar{E}\bar{F})
\end{aligned}
$$

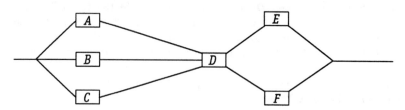

Figure 6.3. Block diagram for Example 6.2

Suppose all units fail with probability γ, independently of each other. Then

$$R = 1 - [\gamma^3 + \gamma + \gamma^2 - \gamma^4 - \gamma^3 - \gamma^5 + \gamma^6]$$
$$= 1 - \gamma - \gamma^2 + \gamma^4 + \gamma^5 - \gamma^6$$

Example 6.3: Figure 6.4 depicts the stages in a journey by public transport from a rural location in southern England to London's Heathrow Airport. Every stage of the journey has some probability of failure, in that some malfunction will prevent the traveller catching her flight. Blocks B, D, E, F have probability of failure p and blocks A, C have probability of failure $2p$, and all blocks fail independently of each other.

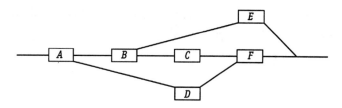

Figure 6.4. Block diagram for Example 6.3

There are just three ties for this system, $\{ADF\}$, $\{ABCF\}$, $\{ABE\}$, but a large number of cuts. It is useful to illustrate the concept of *minimal* cut here. The set $\{BCF\}$ is a cut, but it is not minimal because the subset $\{BF\}$ is also a cut and, further, it is one which cannot be reduced and still be a cut. $\{BF\}$ is minimal and any addition to this is completely redundant. The full list of minimal cuts is $\{A\}$, $\{BF\}$, $\{BD\}$, $\{ECD\}$, $\{EF\}$.

Using ties, the reliability of the system is given by

$$P(T_1 \cup T_2 \cup T_3) = P(T_1) + P(T_2) + P(T_3) - P(T_1 \cap T_2) - P(T_1 \cap T_3)$$
$$- P(T_2 \cap T_3) + P(T_1 \cap T_2 \cap T_3)$$

$$= P(ADF) + P(ABCF) + P(ABE) - P(ABCDF) - P(ABDEF)$$
$$- P(ABCEF) + P(ABCDEF)$$

$$= (1 - 2p)(1 - p)^2 + (1 - 2p)^2(1 - p)^2 + (1 - 2p)(1 - p)^2$$
$$- (1 - 2p)^2(1 - p)^3 - (1 - 2p)(1 - p)^4$$
$$- (1 - 2p)^2(1 - p)^3 + (1 - 2p)^2(1 - p)^4$$

$$= (1 - 2p)(1 - p)^2[1 + 1 - 2p + 1 - 2(1 - 2p) + (1 - p) - (1 - p)^2$$
$$+ (1 - 2p)(1 - p)^2]$$

$$= (1 - 2p)(1 - 2p + p^2)(1 + 2p - 2p^3)$$

$$= 1 - 2p - 3p^2 \dots .$$

If p is very small, such that terms in p^2 are too small to be significant, then

$$R \approx 1 - 2p - 2p + 2p = 1 - 2p.$$

This reflects the fact that A is a critical connection. Once through A the chance of finding a successful path is high.

6.4   Fault trees

The reliability level of a product is established at the design phase. Procedures such as design review, failure mode effects and criticality analysis (FMEA/FMECA) and fault tree analysis (FTA) have been developed for application early in the design phase of a new product.

Failure mode analyses are preliminary design evaluation procedures to identify design weaknesses that may result in safety hazards or reliability problems. It can be regarded as a 'what if ...' approach.

Fault tree analysis begins with the definition of an undesirable event and traces this event down through the system to identify basic causes. A fault tree is a diagram of the event combinations which cause a system to fail. System failure is known as the *top event*. Events are represented by symbols, commonly those shown in Table 6.1. Block diagrams and fault trees are directly equivalent, but complementary in that block diagrams model successful paths and fault trees model paths leading to system failure.

Example 6.4: A fault tree equivalent to the system shown in Figure 6.5 is given in Figure 6.6.

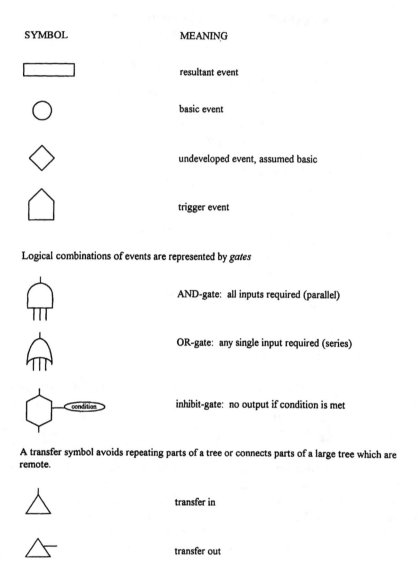

SYMBOL   MEANING

resultant event

basic event

undeveloped event, assumed basic

trigger event

Logical combinations of events are represented by *gates*

AND-gate: all inputs required (parallel)

OR-gate: any single input required (series)

inhibit-gate: no output if condition is met

A transfer symbol avoids repeating parts of a tree or connects parts of a large tree which are remote.

transfer in

transfer out

Table 6.1. Fault tree symbols

If we take the symbol X_i to mean that X_i works, system success is given by $(X_1 \cap X_2) (X_3 \cap X_4)$. System failure is given by $(\overline{X}_1 \cup \overline{X}_2) \cap (\overline{X}_3 \cup \overline{X}_4)$. This is related to system success by taking the complements of basic events and replacing $\cap$ by $\cup$ and vice versa.

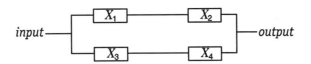

Figure 6.5. Block diagram for Example 6.4

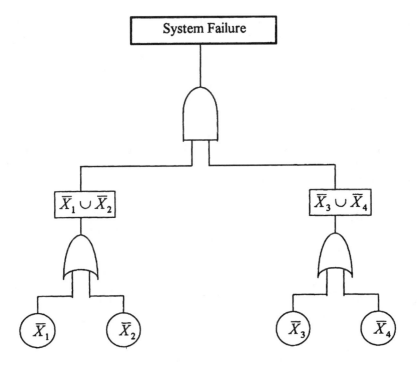

Figure 6.6. Fault tree for Example 6.4

The steps of the analysis can be summarized as follows:

(i)   Split fault tree into smaller trees.
(ii)   Find minimal cuts.
(iii)   Find successful paths.
(iv)   Evaluate reliabilities.
(v)   Identify: (a) weak points; (b) over-design/system simplification; and
        (c) the effect of a given component.

A good introductory guide to failure mode effects and fault tree analysis is given in Leitch (1995), and a number of examples illustrating fault tree analysis can be found in Klaassen and van Peppen (1989).

6.5   Failure over time

The previous sections dealt with the failure of a component part of a system via a 'one-shot' probability, which could, for example, be the probability of failure over a specified period of time, say for an aircraft with a particular flight or 'mission' time. Now we will allow time to be a variable and seek to derive system reliability as a function of time. In most of the foregoing the system components will be described by exponential failure time distributions, but the principles are transferable to other failure mechanisms. However, for complex systems the calculations quickly become unwieldy.

In the reliability block diagrams the blocks will be labelled with the appropriate exponential failure rate parameter. Figure 6.7 shows series and parallel system structures with exponential components having different rate parameters.

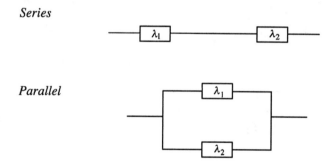

Figure 6.7. System structures whose components have different rate parameters

The series system only works if both units work, so the reliability function

$$R(t) \; = \; e^{-\lambda_1 t} e^{-\lambda_2 t} \; = \; e^{-(\lambda_1 + \lambda_2)t} . \tag{6.4}$$

Thus the failure time of this system is also exponential, rate $(\lambda_1 + \lambda_2)$ and the MTTF is $1/(\lambda_1 + \lambda_2)$.

This simple series arrangement of components with exponential lifetimes is the only arrangement resulting in an exponential system lifetime but it is often common practice for reliability predictions to assume that for any system the failure rate is the sum of the component failure rates. This is the so-called 'parts count' approach which is essentially a 'worst case' scenario description.

In the parallel case, the system only fails if both units fail, so

$$
\begin{aligned}
R(t) &= 1 - \bar{R}(t) \\
&= 1 - (1 - e^{-\lambda_1 t})(1 - e^{-\lambda_2 t}) \\
&= e^{-\lambda_1 t} + e^{-\lambda_2 t} - e^{-(\lambda_1 + \lambda_2)t}.
\end{aligned}
\tag{6.5}
$$

The hazard function, $h(t)$, is now time-dependent:

$$
h(t) = \frac{-R'(t)}{R(t)} = \frac{\lambda_1 e^{-\lambda_1 t} + \lambda_2 e^{-\lambda_2 t} - (\lambda_1 + \lambda_2)e^{-(\lambda_1 + \lambda_2)t}}{e^{-\lambda_1 t} + e^{-\lambda_2 t} - e^{-(\lambda_1 + \lambda_2)t}}.
$$

It is informative to look at the case where $\lambda_1 = \lambda_2$. Then

$$
h(t) = \frac{2\lambda e^{-\lambda t} - 2\lambda e^{-2\lambda t}}{2e^{-\lambda t} - e^{-2\lambda r}} = \frac{\lambda(1 - e^{1\lambda t})}{1 - \frac{1}{2}e^{-\lambda t}}.
$$

The hazard is zero at time $t = 0$ and increases over time, ultimately approaching λ.

The MTTF in the general case is given by

$$
\frac{1}{\lambda_1} + \frac{1}{\lambda_2} - \frac{1}{\lambda_1 + \lambda_2}.
$$

For two parallel units with the same failure rate λ,

$$
\begin{aligned}
\text{MTTF} &= \frac{1}{\lambda} + \frac{1}{\lambda} - \frac{1}{2\lambda} = \frac{3}{2\lambda} \\
&= \frac{1}{\lambda}\left(1 + \frac{1}{2}\right).
\end{aligned}
$$

For n parallel units with the same failure rate λ,

$$\text{MTTF} = \frac{1}{\lambda}\left(1 + \frac{1}{2} + \frac{1}{3} + \dots + \frac{1}{n}\right).$$

Thus the diminishing benefit of adding successive parallel units can be seen.

6.6   Quorum or m-out-of-n systems

Here the system functions if at least m units function. For $m = n$ the system is a pure series system and for $m = 1$ it is pure parallel.

If we assume that units are independent, each with reliability p, then the number of units working is a random variable, X say, having a *binomial distribution* with parameters (n, p).

The probability that k out of n units work is given by

$$P(X = k) = \binom{n}{k} p^k (1 - p)^{n-k}, \tag{6.6}$$

where

$$\binom{n}{k} = \frac{n!}{k!(n-k)!}.$$

This is the probability that k units work and $n - k$ units fail times the number of ways that there can be only k of the n units working.

An m-out-of-n system is reliable if m, $m + 1$, $m + 2$, ..., n units function correctly. Therefore

$$R(t) = \sum_{k=m}^{n} \binom{n}{k} p^k (1 - p)^{n-k}. \tag{6.7}$$

For n active units with exponential failure time, rate λ,

$$R(t) = \sum_{k=m}^{n} \binom{n}{k} (e^{-\lambda t})^k (1 - e^{-\lambda t})^{n-k}$$

and

$$\text{MTTF} = \frac{1}{\lambda} \sum_{k=m}^{n} \frac{1}{k}.$$

Example 6.5: The reliability of a two-out-of-three system with similar units is

$$R(t) = 3p^2(1 - p) + p^3 = 3p^2 - 2p^3.$$

If the units have exponential lifetime, $p = e^{-\lambda t}$, then

$$R(t) = 3e^{-2\lambda t} - 2e^{-3\lambda t}$$

and

$$\text{MTTF} = \frac{1}{\lambda}\left(\frac{1}{2} + \frac{1}{3}\right) = \frac{5}{6\lambda}.$$

If the lifetimes are Weibull distributed, $p = e^{-(t/\alpha)^\beta}$, then

$$R(t) = 3e^{-(t/\alpha)^{2\beta}} - 2e^{-(t/\alpha)^{3\beta}}.$$

6.7 Redundancy

Redundancy may take two essential forms, *active* or *passive*. Active redundancy is when all parts of a system operate continuously but the system may still function if only some of the components work. For example, an aircraft with four engines may still be able to fly if only two function. It may be that in such systems components continuing to function after others have failed will be under increased load which could affect their failure rates.

Passive or *standby* redundancy is where components are installed which are surplus to requirements and are only activated when certain other components fail. Redundancy involves some form of parallel design in the system.

Both types of redundancy lead to increased reliability, but in general a greater improvement is achieved with standby than with active redundancy since standby units operate for less time than active ones. An option

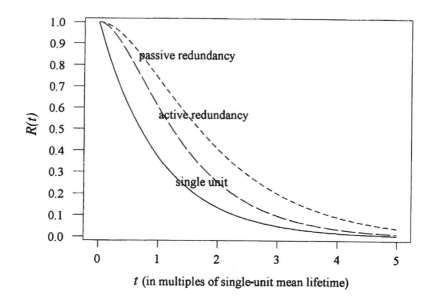

t (in multiples of single-unit mean lifetime)

Figure 6.8. Reliability function for single-unit and two-unit redundant systems (all units with same exponential lifetime model)

exists in some applications to have *partially active* redundancy, whereby units just 'tick over' while waiting to be called upon, rather than being completely switched off. This may be to avoid a 'warming-up period' or to assist the speed of engaging the unit. In practice, redundancy does not simply involve the addition of units. It usually means that switching mechanisms are more complex. Figure 6.8 illustrates the general effect of redundancy in a system of two units with exponential lifetimes.

The two-unit active redundant system has an MTTF of $3/2\lambda$, and a two-unit passive redundant system has an MTTF of $2/\lambda$.

Example 6.6: A classic question is the comparison of (i) redundancy at system level with (ii) redundancy at component level. A system of two main units and two active standby units will be examined here. Figure 6.9 shows the respective RBDs.

Calculating the reliability of both systems involves application of (6.4) and (6.5) but in different orders. For (i)

$$R(t) = 1 - [1 - R_1(t)R_2(t)]^2 = 2R_1(t)R_2(t) - [R_1(t)R_2(t)]$$

$$= R_1R_2[2 - R_1R_2].$$

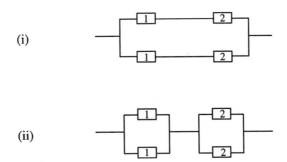

Figure 6.9. Block diagrams for Example 6.6

For (ii),

$$R(t) = [1 - (1 - R_1(t))^2][1 - (1 - R_2(t))^2]$$

$$= 2R_1(t)R_2(t)[2 - R_1(t) - R_2(t)] + [R_1(t)R_2(t)]^2$$

$$= R_1R_2[2 - R_1][2 - R_2].$$

Figures 6.10 and 6.11 show these two system reliabilities for exponential units with parameters $\lambda_1 = 0.0005$ and $\lambda_2 = 0.001$, and for units having Weibull distributed lifetimes, both with shape parameter 5 and scale parameter 1060. In each case the 'mission' time is 1000.

System (ii) is always more reliable than system (i) but this will in practice usually be subject to the reliability of the extra failure sensing/switching mechanism required in (ii). This could be modelled by using a slightly higher hazard rate for unit 2 in system (ii) than in system (i). Alternative approaches are considered in Section 6.10.

6.8   Analysis of systems using state spaces

A closer analysis of what actually occurs in the failure of the systems described above is that the system moves from state to state, one component failing after another until the last component fails. This is an example of state *transition*.

Consider a two-unit parallel system with standby redundancy, each unit having failure rate λ. Possible system states are: S_0, unit 1 in operation; S_1, unit 2 in operation; and S_2, system failed. Using the hazard function definition of λ, the probability of passing from

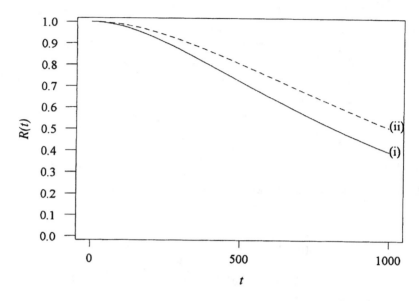

Figure 6.10. Comparison of redundancy at (i) system level (ii) component level for Example 6.6 exponential lifetimes

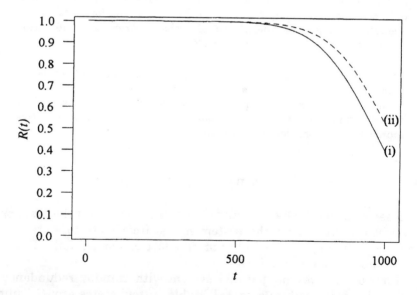

Figure 6.11. Comparison of redundancy at (i) system level (ii) component level for Example 6.6 unit Weibull distributed lifetimes

state S_0 to S_1 in any given time interval $[t, t+\delta t]$, given that the current state is S_0, is $\lambda\delta t$. Similarly, conditional on being in state S_1 at time t, the probability of passing to S_2 in the next δt is $\lambda\delta t$. The probabilities of not moving from S_0 or S_1 follow, and since there is no repair in this situation, once arrived at state S_2, the system remains in S_2. This probabilistic description of the possible transitions can be illustrated by a *Markov diagram*, shown for the above system in Figure 6.12.

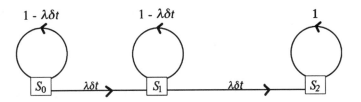

Figure 6.12. Markov diagram for a two-unit parallel system with standby redundancy, each unit with exponentially distributed lifetime

A *Markov system* is one where the transition probabilities only depend on the *present state* of the system. Markov analysis requires:

(i) a matrix of transition probabilities describing the probability of moving from state i to state j, given that a move is being made; and

(ii) the 'waiting time' distribution for each state, until the move is made.

The analysis is much easier when we can assume that (ii) is exponential, rate λ_i for state i, depending only on i. Let $P_i(t)$ be the probability of being in state i at time t.

The *transition matrix* shows the probabilities of passing from one state to another during the time interval $[t, t + \delta t]$:

$$
\begin{array}{c}
\text{current state} \\
\begin{array}{ccc}
0 & 1 & 2
\end{array}
\end{array}
$$

$$
\text{next state}\;
\begin{array}{c} 0 \\ 1 \\ 2 \end{array}
\left[
\begin{array}{ccc}
1 - \lambda\delta t & 0 & 0 \\
\lambda\delta t & (1 - \lambda\delta t) & 0 \\
0 & \lambda\delta t & 1
\end{array}
\right] .
$$

The probability of being in state i at time $[t+\delta t]$ is $P_i(t + \delta t)$ and is given by applying formula (1.5). This can be expressed in matrix form as follows:

$$\begin{bmatrix} 1 - \lambda \delta t & 0 & 0 \\ \lambda \delta t & 1 - \lambda \delta t & 0 \\ 0 & \lambda \delta t & 1 \end{bmatrix} \begin{bmatrix} P_0(t) \\ P_1(t) \\ P_2(t) \end{bmatrix} = \begin{bmatrix} P_0(t + \delta t) \\ P_1(t + \delta t) \\ P_2(t + \delta t) \end{bmatrix}.$$

Multiplying out and letting $\delta t \to 0$ yields a set of differential equations:

$$\begin{aligned} P_0'(t) &= -\lambda P_0(t), \\ P_1'(t) &= -\lambda P_0(t) - \lambda P_1(t), \\ P_2'(t) &= \qquad\quad \lambda P_1(t). \end{aligned} \qquad (6.8)$$

The first equation, for example, is from

$$(1 - \lambda \delta t)P_0(t) = P_0(t + \delta t)$$
$$-\lambda P_0(t) = \frac{P_0(t + \delta t) - P_0(t)}{\delta t}.$$

By definition, the right-hand side is the derivative of $P_0(t)$ when δt is allowed to go to zero. In other words the differential equations (sometimes referred to as *Chapman–Kolmogorov* equations) describe the instantaneous movement of the system, or the movement over continuous time.

There are standard ways of solving these equations. One structured way is offered by the use of Laplace transforms. Recalling (5.4), it is possible to express the equations in terms of $P_i^*(s)$ only. If we assume that the system is in state 0 at time zero, we have

$$\begin{aligned} sP_0^*(s) - 1 &= -\lambda P_0^*(s) \\ sP_1^*(s) &= -\lambda P_0^*(s) - \lambda P_1^*(s), \\ sP_2^*(s) &= \qquad\quad -\lambda P_1^*(s). \end{aligned}$$

Hence,

$$P_0^*(s) = \frac{1}{s + \lambda},$$

$$P_1^*(s) = \frac{\lambda}{s + \lambda} P_0^*(s).$$

One of the equations is always redundant because by definition the $P_i(t)$ sum to one. The transforms can be inverted by reference to Table 5.2:

$$P_0(t) = e^{-\lambda t},$$

$$P_1^*(s) = \frac{\lambda}{(s + \lambda)} \Rightarrow P_1(t) = \lambda t e^{-\lambda t}.$$

So the reliability of the system is

$$P_0(t) + P_1(t) = e^{-\lambda t}(1 + \lambda t).$$

This is a special case of the general result deduced in Example 2.3. The state space analysis is the more complicated approach for this particular system, but serves to demonstrate the method.

Example 6.7: Consider a standby redundant system where the standby is partially active. Both components have constant failure rate λ_1 when operating as the main component, but the standby runs in partially active mode while waiting to replace the main component when it fails and in this state has failure rate λ_2 ($< \lambda_1$). The states of the system can be summarized as follows:

	unit 1	unit 2
S_0	main	available
S_1	main	failed
S_2	failed	main
S_3	failed	failed

The Markov diagram, shown for simplicity with just the transition rates, is given in Figure 6.13(a).

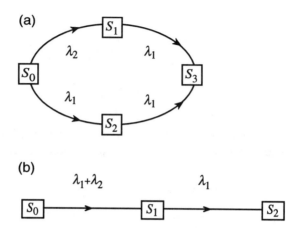

Figure 6.13. (a) Markov diagram for a two-unit system with partially active redundancy (b) Simplified representation of Figure 6.13(a)

Now while this is a perfectly reasonable description of the situation, it is not the simplest because we can describe the system behaviour using fewer states. Minimizing the number of states leads to a simpler diagrammatic representation of a system and, most importantly, to a smaller set of differential equations to deal with. Suppose we consider the system from the point of view of how many working units are available. In principle, there is no operational difference between states 1 and 2 above so there is little benefit in distinguishing between them. What would make those states different is different failure rates thereafter. A simplified Markov diagram is shown in Figure 6.13(b), where state S_i means $2 - i$ units available.

Making use of the similarity in this description and that of the two-unit passive system, the system of differential equations yielded by the analysis is

$$
\begin{aligned}
P_0'(t) &= -(\lambda_1 + \lambda_2)P_0(t), \\
P_1'(t) &= (\lambda_1 + \lambda_2)P_0(t) - \lambda_1 P_1(t), \\
P_2'(t) &= \qquad\qquad\qquad \lambda_1 P_1(t).
\end{aligned}
\tag{6.9}
$$

So

$$
P_0^*(s) = \frac{1}{s + \lambda_1 + \lambda_2},
$$

$$P_1^*(s) = \frac{\lambda_1 + \lambda_2}{(s + \lambda_1)(s + \lambda_1 + \lambda_2)},$$

given that the system is in state S_0 at time $t = 0$. It follows from Table 5.2 that

$$P_0(t) = e^{-(\lambda_1 + \lambda_2)t} \ldots$$

$$P_1(t) = \frac{\lambda_1 + \lambda_2}{\lambda_2}(e^{-\lambda_1 t} - e^{-(\lambda_1 + \lambda_2)t})$$

and

$$R(t) = \frac{1}{\lambda_2}[(\lambda_1 + \lambda_2)e^{-\lambda_1 t} - \lambda_1 e^{-(\lambda_1 + \lambda_2)t}].$$

Were this system operating as a fully active parallel system, that is $\lambda_2 = \lambda_1$, then $R(t)$ would reduce to the form of (6.5) with $\lambda_2 = \lambda_1$. However, it should be noted that it is not possible to let $\lambda_2 = 0$ in $R(t)$ in order to obtain the reliability of a completely passive system because $P_1^*(s)$ takes a different form from an inversion point of view.

6.9   Mean time to failure (MTTF)

A component or system with exponential lifetime, rate parameter λ, has MTTF $1/\lambda$. A system described by a Markov diagram with only one route through the diagram and constant transition rates (Figure 6.14) has MTTF consisting of the sum of mean times spent in each state:

$$\text{MTTF} = \frac{1}{\lambda_0} + \frac{1}{\lambda_1} + \frac{1}{\lambda_2} \ldots .$$

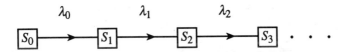

Figure 6.14. Markov diagram for a system with only one path

For systems not so easily specified, the MTTF is obtained from (1.8), but more simply from $R^*(s)$, the sum of the Laplace transforms of the $P_i(t)$ for working system states:

$$\text{MTTF} = R^*(0) \tag{6.5}$$

For Example 6.7,

$$R^*(s) = P_0^*(s) + P_1^*(s) = \frac{1}{s + \lambda_1 + \lambda_2} + \frac{\lambda_1 + \lambda_2}{(s + \lambda_1)(s + \lambda_1 + \lambda_2)}.$$

So,

$$\text{MTTF} = \frac{1}{\lambda_1 + \lambda_2} + \frac{\lambda_1 + \lambda_2}{\lambda_1(\lambda_1 + \lambda_2)} = \frac{1}{\lambda_1 + \lambda_2} + \frac{1}{\lambda_1},$$

which would follow from the simplified Markov diagram.

6.10   Considerations due to 'switching'

Reliability gains as a result of redundancy are almost always partly offset by the resulting increased complexity of the system design. The additional units are in practice accompanied by devices which detect failure, which 'switch' in the units or which monitor output in quorum or 'voting' systems. These peripheral features must be highly reliable if they are to support the redundancy.

> **Example 6.8:** An electronic device has a main unit which when it fails is repaired, and meanwhile a standby unit is switched in to keep the device functioning. The switching mechanism has a probability, p, of working when called upon and the units have exponential lifetimes with rate parameter λ. This is the same system described in Section 6.8 but the transition from S_0 to S_1 is dependent also on the switch working and there is now the possibility of passing from S_0 to S_2 as a result of the switch not working. The Markov diagram now becomes that shown in Figure 6.15.

The differential equations (6.8) become

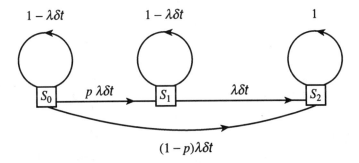

Figure 6.15. Markov diagram for Example 6.8

$$P_0'(t) = -\lambda P_0(t),$$
$$P_1'(t) = p\lambda P_0(t) - \lambda P_1(t),$$
$$P_2'(t) = (1-p)\lambda P_0(t) + \lambda P_1(t),$$

and the reliability of the system is $R(t) = e^{-\lambda t}(1 + p\lambda t)$.

A value of p which is not close to 1 will have a severe impact on the benefits of redundancy.

Example 6.8 (continued): To give an example of a practical consideration, suppose the main unit takes 3 hours to repair. 'Down' time will be minimized if there is a high probability that the standby system works until the main unit is restored. We will let the value of p be 0.95 and give the standby unit a mean lifetime of 50 hours. The probability that the standby system fails while the main unit is under repair is

$$P_f = P(\text{switch works and standby lifetime} < 3 \text{ hours}) + P(\text{switch fails})$$

$$= 0.95\left[1 - \exp\left(-\frac{3}{50}\right)\right] + 0.05 = 0.1053.$$

Now suppose that for the same cost, either a switch with reliability 0.99 or a standby unit with mean lifetime 100 hours (exponential lifetime again) could be used. The resulting probability of system failure under these options is first,

$$P_f = 0.99\left[1 - \exp\left(-\frac{3}{50}\right)\right] + 0.01 = 0.0676,$$

and second,

$$P_f = 0.95\left[1 - \exp\left(-\frac{3}{100}\right)\right] + 0.05 = 0.0780.$$

The more reliable switch is therefore the better option. A different model for the unit lifetime will lead to different numerical comparisons but the general conclusion about switching mechanisms will remain the same.

Example 6.9: Monitoring devices may be in general more appropriately modelled as for the main units, by time-dependent lifetime models. Suppose we consider a 'failure detection' device acting as controller in a two-unit standby redundant system. Let the units have failure rate parameter λ and the controller fail to detect main unit failure and activate the standby with rate λ_S. In effect, the controller acts in series with the main unit and therefore their failure rates may be added. The Markov diagram reduces to that of Example 6.7 with transition rates $\lambda + \lambda_s$ and λ. The differential equations follow the form of (6.9) and the resulting reliability is

$$R(t) = \frac{1}{\lambda_s}[(\lambda + \lambda_s)e^{-\lambda t} - \lambda e^{-(\lambda + \lambda_s)t}].$$

Passive redundancy necessarily involves more control than active redundancy so it is worth examining how far this control reduces the advantage of passive mode. Assuming that there is negligible additional failure possibility in active mode, the Markov diagram is Figure 6.16(a), and the passive case is shown in Figure 6.16(b).

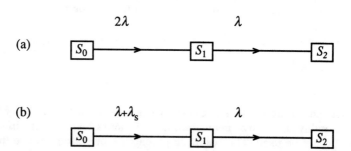

Figure 6.16. Markov diagrams for Example 6.9

The passive system is more reliable if $\lambda_s < \lambda$. Clearly, though, this condition needs to be comfortably satisfied given that there are likely to be factors such as cost and size to be taken into account.

Example: 6.10: A further complication may be that the controller detects a failure when there is none. Again this event has the effect of reducing the system from two working units to one, this time the standby. So if this event occurred say at rate λ_0 the Markov diagram would be as above with the transition rate from S_0 to S_1 now $\lambda + \lambda_s + \lambda_0$.

6.11   Common cause failures

A system may be subject to a cause of failure which affects a whole group of components at the same time. A typical example is a general power failure. Usually it is some kind of influence which is external to the units. We shall not deal here with failure which is caused by the failure of other units. Failure dependency is discussed in Chapter 10.

Example 6.11: A computer-controlled process has two computers, each able to run the process on its own. Both computers are connected to a central power supply. The computers and power supply have exponential time to failure, rates λ_1, λ_2 and λ_c respectively, where λ_1 and λ_2 do not incorporate failures due to failure of the power supply. Each computer is subject to its own failure and that of the power supply, but the common cause failure will result in both units failing. In effect the power supply acts in series with the computer system, as shown in Figure 6.17. Combining (6.4) with (6.5), the reliability of the system is given by

$$R(t) = e^{-\lambda_c t}[e^{-\lambda_1 t} + e^{-\lambda_2 t} - e^{-(\lambda_1 + \lambda_2)t}].$$

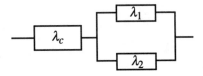

Figure 6.17. A block diagram depicting common cause failure

CHAPTER 7

Models for Functions of Random Variables

7.1 Combinations and transformations of random variables

The reliability of a system is usually a function of several variables and parameters, most of which are random variables. In the case of large complex systems, collection of data sufficient to give a complete characterization of performance may not be possible. However, if the total system can be broken down into subsystems and components, and the statistical behaviour of individual parts studied, then this may be integrated into a complete description of the behaviour of the system.

Given $Y = f(X_1, X_2, ..., X_n)$, how do we find certain properties of Y as functions of the properties of the random variables $\{X_i\}$? The variables $\{X_i\}$ may or may not be independent.

7.2 Expectations of functions of random variables

Let X and Y denote two random variables with expectations $E(X) = \mu_X$ and $E(Y) = \mu_Y$, respectively, and a denote a constant. Then

$$
\begin{aligned}
E(aX) &= aE(X), \\
E(a + X) &= a + E(X), \\
E(X \pm Y) &= E(X) \pm E(Y).
\end{aligned}
\tag{7.1}
$$

If X and Y are independent,

$$E(XY) = E(X)\,E(Y). \tag{7.2}$$

Denote by σ_X^2 and σ_Y^2 the variances $V(X)$ and $V(Y)$, respectively. Then

$$V(X) = E[X - E(X)]^2 = E(X^2) - [E(X)]^2.$$

Therefore,

$$E(X^2) = [E(X)]^2 + V(X)$$

and

$$
\begin{aligned}
V(aX) &= a^2 V(X) \\
V(a + X) &= V(X) \\
V(X^2) &= E(X^4) - (\mu_X^2 + \sigma_X^2)^2
\end{aligned}
\tag{7.3}
$$

If X and Y are independent,

$$V(X \pm Y) = V(X) + V(Y) \tag{7.4}$$

and

$$V(XY) = \sigma_X^2 \sigma_Y^2 + \sigma_Y^2 \mu_X^2 + \sigma_X^2 \mu_Y^2. \tag{7.5}$$

Example 7.1: A rectangular steel plate has width X and length Y, both dimensions being independent random variables. The width has mean 20 centimetres and standard deviation 2 millimetres. The length has mean 40 centimetres and standard deviation 2 millimetres. The area of the plate is say $XY = Z$.

From (7.2),

$$E(Z) = 40 \times 20 \text{ cm}^2.$$

From (7.5)

$$V(Z) = 0.2^2 \times 0.2^2 + 0.2^2 \times 20^2 + 0.2^2 \times 40^2 \text{ cm}^4.$$

So the standard deviation of Z is 8.944 cm^2.

If X and Y are dependent, then (7.1) still applies but

$$V(X \pm Y) = V(X) + V(Y) \pm 2C(X, Y), \qquad (7.6)$$

where $C(X,Y)$ is the *covariance* of X and Y, defined by

$$C(X,Y) = E([X - E(X)][Y - E(Y)]) = E(XY) - E(X)E(Y).$$

If Y tends to increase as X increases then $C(X,Y)$ is positive. If Y tends to decrease as X increases then $C(X,Y)$ is negative. If X and Y are statistically independent $C(X,Y)$ is zero and (7.6) yields (7.4).

7.3   Approximations for $E[g(X)]$ and $V[g(X)]$

In general it is quite difficult to find the density function for a function of random variables. Knowledge of the moments of the transformed variable can be quite useful.

Probability bounds can be obtained using *Chebyshev's inequality*, given by

$$P(|Y - \mu| \geq k\sigma) \leq \frac{1}{k^2},$$

where Y is a random variable with mean μ and variance σ^2, or *Gauss' inequality*,

$$P(|Y - \mu| \geq k\sigma) \leq \frac{4}{9k^2},$$

if the density of Y has only one maximum, Y_{max}, and if $E(Y) = Y_{max}$, but these tend to result in very generous bounds. For example, the normal distribution has approximately 5% of its distribution outside two standard deviations from the mean. For the case $k = 2$ the above inequalities would yield probability boundaries 25% and 11%, respectively.

Consider the case where X is a one-dimensional variable. A Taylor expansion (see the Appendix) of $Y = g(X)$ about the point $X = \mu$ is given by

$$Y = g(\mu) + (X - \mu)g'(\mu) + \frac{(X - \mu)^2}{2!} g''(\mu) + \Re$$

where $\Re$ is the remainder. Taking expectations,

$$E(Y) = E[g(\mu)] + E[Xg'(\mu) - \mu g'(\mu)] + E\left\{\frac{1}{2}g''(\mu)(X - \mu)^2\right\} + E(\Re)$$

$$= g(\mu) + \{\mu g'(\mu) - \mu g'(\mu)\} + \frac{1}{2}g''(\mu)V(X) + E(\Re) \qquad (7.7)$$

$$\approx g(\mu) + \frac{1}{2}g''(\mu)V(X)$$

We can also form the Taylor expansion of $V(Y)$ to two terms to give

$$V(Y) \approx V[g(\mu)] + V[(X - \mu)g'(\mu)]$$
$$= [g'(\mu)]^2 V(X) \qquad (7.8)$$

Example 7.2: If X has a normal distribution, mean μ and variance σ^2, then $Y = e^X$ has a lognormal distribution. The function $g(x)$ is e^x and so $g'(x)$ and $g''(x)$ are also e^x. Expression (7.3) yields

$$E(Y) = e^\mu + \frac{1}{2}e^\mu \sigma^2 = e^\mu\left(1 + \frac{1}{2}\sigma^2\right).$$

The true $E(Y)$ is

$$e^{\mu + \frac{1}{2}\sigma^2} = e^\mu e^{\frac{1}{2}\sigma^2}.$$

For small α, e^α is approximately $1 + \alpha$, so for small σ^2 expression (7.7) gives a good approximation. Similarly, (7.8), which yields

$$V(Y) \approx (e^\mu)^2 \sigma^2,$$

is a close approximation, under the same condition, to the true $V(Y)$, which is

$$e^{2\mu + \sigma^2}(e^{\sigma^2} - 1).$$

These results may be extended to the case where g is a function of n independent random variables to give

$$E(Y) \approx g(\mu_1, \mu_2, \dots, \mu_n), \qquad (7.9)$$

$$V(Y) \approx \sum_{i=1}^{n} \left\{ \frac{\partial g(\mathbf{X})}{\partial X_i} \bigg| \mathbf{X} = \mathbf{\mu} \right\}^2 V(X_i).$$ (7.10)

Example 7.3: For Example 7.1, expression (7.9) would yield $E(Z) = \mu_X \mu_Y$, which is exact. Using (7.10),

$$V(Z) = \mu_Y^2 \sigma_X^2 + \mu_X^2 \sigma_Y^2.$$

Given the relatively small standard deviations of X and Y this expression proves to be a very close approximation to the true value.

Example 7.4: A tolerance interval for a random variable is often given in the form $\mu \pm k\sigma$. The tolerances for the sides X, Y, Z of a rectangular solid bar are given, using $k = 2$, as: X, 2 ± 0.002 metres; Y, 1 ± 0.001 metres; and Z, 4 ± 0.008 metres. We conclude therefore that $\mu_X = 2$, $\sigma_X = 0.001$, $\mu_Y = 1$, $\sigma_Y = 0.0005$, $\mu_Z = 4$, $\sigma_Z = 0.004$. Assuming X, Y and Z to be independent of each other, formula (7.5) may be extended to give the variance of the volume XYZ of the bar:

$$V(XYZ) = \sigma_X^2 \sigma_Y^2 \sigma_Z^2 + \sigma_X^2 \sigma_Y^2 \mu_Z^2 + \sigma_X^2 \mu_Y^2 \sigma_Z^2 + \mu_X^2 \sigma_Y^2 \sigma_Z^2 + \mu_X^2 \mu_Y^2 \sigma_Z^2$$
$$+ \mu_X^2 \sigma_Y^2 \mu_Z^2 + \sigma_X^2 \mu_Y^2 \mu_Z^2.$$

The approximation (7.10) gives only the last three terms of this series (0.000096) but for the means and standard deviations specified by the tolerance intervals all other terms are relatively very small. As in Example 7.3, the exact and approximate formulae give the same expression for the mean volume, $\mu_X \mu_Y \mu_Z = 8$. A tolerance interval for the volume of the bar is therefore

$$8 \pm 2 \sqrt{0.000096} = 8 \pm 0.0196.$$

In general, exact forms for the mean and variance of a function of random variables will be hard to derive but the approximations (7.9) and (7.10) will be good alternatives, particularly where the variances of component variables are small.

7.4 Distribution of a function of random variables

Consider the case of one random variable; given $Y = u(X)$ we wish to find the density function $f_Y(y)$ for the random variable Y, assuming that we know the density function $f_X(x)$ of the random variable X.

We can show that

$$f_Y(y) = \left|\frac{dk}{dy}\right| f_X(k(y)),$$

where

$$k(y) = u^{-1}(y) = x.$$

The term $|dk/dy|$ represents the absolute value of the derivative of x with respect to y.

If the function $k(y)$ is double-valued there will be two values of x for each y. Say these are denoted x_1 and x_2; then

$$f_Y(y) = \left|\frac{dx_1}{dy}\right| f_X(x_1) + \left|\frac{dx_2}{dy}\right| f_X(x_2).$$

In general, if there are n possible values of $k(y)$, x_1, x_2, ..., x_n, there will be n terms contributing to $f_Y(y)$.

Example 7.5: Consider a circular bar of diameter d. To compute stress, the cross-sectional area $a = (\pi/4)d^2$ is required. Let diameter be a random variable, D, having a normal distribution, mean μ and variance σ^2. Now

$$d = u^{-1}(a) = \pm\sqrt{4a/\pi}.$$

Differentiating,

$$\left|\frac{dd}{da}\right| = \sqrt{1/a\pi},$$

so

$$f_A(a) = \sqrt{1/a\pi}[f_D(\sqrt{4a/\pi}) + f_D(-\sqrt{4a/\pi})]$$

$$= \frac{\sqrt{1/a\pi}}{\sigma\sqrt{2\pi}}\left\{\exp\left[-\frac{(\sqrt{4a/\pi}-\mu)^2}{2\sigma^2}\right] + \exp\left[-\frac{(-\sqrt{4a/\pi}-\mu)^2}{2\sigma^2}\right]\right\}.$$

In practice, the contribution from $d = -\sqrt{4a/\pi}$ will be negligible. Clearly d cannot be negative and while its normal distribution allows for negative values, the probability in the tail of the distribution to the left of $d = 0$ will be virtually zero.

The procedure extends in a general way to the case where Y is a function of a *vector* $\mathbf{X}$. Let $f_{\mathbf{X}}(\mathbf{x})$ be the joint probability density function of the vector $\mathbf{X}$ and the function $\mathbf{u}$ be a one–one mapping $\mathbf{x} \rightarrow \mathbf{y}$ with $\mathbf{k} = (k_1, k_2) = \mathbf{u}^{-1}$. We now have that

$$f_{\mathbf{Y}}(\mathbf{y}) = \mid J(\mathbf{y}) \mid f_{\mathbf{X}}(\mathbf{k}(\mathbf{y})),$$

where $J(\mathbf{y})$ is the *Jacobian* of the transformation $\mathbf{k}$. $J(\mathbf{y})$ is the determinant of a matrix of partial derivatives. For the bivariate case,

$$J(\mathbf{y}) = \begin{vmatrix} \dfrac{\partial k_1}{\partial y_2} & \dfrac{\partial k_1}{\partial k_2} \\ \dfrac{\partial k_2}{\partial y_1} & \dfrac{\partial k_2}{\partial y_2} \end{vmatrix} = \frac{\partial k_1 \partial k_2}{\partial y_1 \partial y_2} - \frac{\partial k_1 \partial k_2}{\partial y_2 \partial y_1}.$$

Example 7.10 in Section 7.9 illustrates this methodology.

7.5   Probabilistic engineering design

In many complex systems, serious consequences can result from the failure of a single component. In choosing the best structural and mechanical designs, engineers need to take into account such factors as reliability, cost, weight and volume, and a sound assessment of component reliability is an essential requirement at an early stage in the design process.

Fundamentally, a component will have a certain stress-resisting capacity, and if the stress induced by the operating conditions exceeds this capacity, failure will result. Conventionally, designers have relied much on arbitrary multipliers such as safety factors and safety margins, in the belief that failure can be eliminated if these factors are large enough. This can never be the case and use of these factors alone does not give a real indication of how likely the component is to fail. Use of a safety factor is only justified when its value is based on considerable experience with components similiar to that under consideration.

A fact often ignored is that design variables and parameters are generally random variables. In probabilistic design, the design variables and parameters are identified explicitly, and stress and strength distributions determined. From these, component reliabilities can be expressed as functions of the stress and strength distributions. Further, expressions for bounds on reliability can be determined and these can assist in studying design reliability as a function of stress and strength variability.

Probabilistic design has its origins in the aerospace industry. Basic steps in the process are:

(i)   Formulation of an initial design.

(ii)   Estimation of external forces and their probability density functions.

(iii) Analysis of the initial system, and design criteria.

(iv) Material selection using physical, mechanical and economic criteria.

(v)   Determination of the probability density functions of the material characteristics.

(vi) Estimation of strength and failure characteristics of components as functions of engineering and geometric properties, and expected operating conditions.

(vii) Determination of system strength and failure characteristics.

Several iterations may be required to obtain a design with optimum properties.

7.6   Stress and strength distributions

In assessing distributions of *stress* (force per unit area) and *strength* (the stress at which failure occurs), the following factors may apply:

(i)   Use may have to be made of adjustment or degradation factors to allow for the differences between components tested under experimental conditions and those operating under normal conditions in the field.

(ii)   It may be assumed that the strength of components is determined by the weakest point, leading to an extreme value distribution, or alternatively that weaker points receive support from stronger points around them, resulting in an averaging process. In the latter, strength is then related to the mean value of samples from the distribution of strength of all points, leading to a normal distribution. There are, however, some drawbacks to the normal distribution in relation to negative values and its symmetrical nature. Some strength properties, such as fatigue life, may have a lognormal distribution.

(iii) Stress distributions cannot usually be generalized as conveniently as strength distributions. Some loads may be heavily asymmetrically distributed and/or possess wide scatter relative to, say, the normal distribution.

7.7   Interference theory and reliability computations

An expression for the probability that a component, subsystem or system fails when the stress exceeds the strength, will now be derived in the case when the probability density functions for stress and strength are known. The random variable S will represent stress, with probability density function $f_S(s)$; and the random variable X will represent strength with probability density function $f_X(x)$.

The shaded portion in Figure 7.1 shows the *interference region*, which is indicative of the probability of failure. Taking a small interval of width δs along the X,S axis, the probability of a stress value lying in this interval is equal to the area of the interference region covering δs:

$$P\left(s_0 - \frac{\delta s}{2} \leq S \leq s_0 + \frac{\delta s}{2}\right) = f_S(s_0)\,\delta s\,.$$

The probability that the strength x is greater than a certain stress s_0 is given by

$$P(X > s_0) = \int_{s_0}^{\infty} f_X(x)\mathrm{d}x\,.$$

Given that stress and strength are independent, the probability of the stress value lying in the small interval δs and the strength exceeding the stress by this small interval δs is

$$f_S(s_0)\,\delta s \cdot \int_{s_0}^{\infty} f_X(x)\mathrm{d}x\,.$$

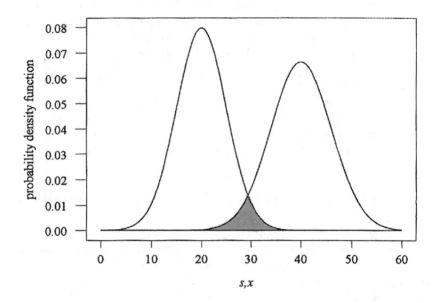

Figure 7.1. Interference region

The reliability of the item is given by summing this probability over the possible values of s_0. Letting $\delta s \to 0$ yields the integral

$$R = P(X > S) = \int_{-\infty}^{\infty} f_S(s)\left[\int_s^{\infty} f_X(x)\mathrm{d}x\right]\mathrm{d}s . \qquad (7.11)$$

Reliability can also be computed on the basis that the stress remains less than the strength. The probability that strength x is within a small interval $(x_0 - \delta x/2, x_0 + \delta x/2)$ is

$$P\left(x_0 - \frac{\delta x}{2} \le X \le x_0 + \frac{\delta x}{2}\right) = f_X(x_0)\delta x$$

and the probability of stress being less than x_0 is given by $\int_{-\infty}^{x} f_S(s)\mathrm{d}s$.

Again assuming S and X to be independent, the probability of the strength belonging to the small interval δx and S not exceeding x_0 is

$$f_X(x)\,\delta x. \int_{-\infty}^{x} f_S(s)\,ds\,.$$

Hence the reliability of the component for all possible values of the strength is

$$R = \int_{-\infty}^{\infty} f_X(x)\left[\int_{-\infty}^{x} f_S(s)\,ds\right]dx\,. \tag{7.12}$$

The use of infinite limits on the integrals is for generality. They may not be practical limits, but will apply where probability distributions are defined over infinite regions, such as for the normal distribution.

Example 7.6: A component with normally distributed strength, mean μ and variance σ^2, and applied stress exponentially distributed, parameter λ has interference region shown in Figure 7.2. Using (7.12),

$$R = \int_{-\infty}^{\infty} \frac{1}{\sigma\sqrt{2\pi}}\exp\left[-\frac{1}{2}\left\{\frac{x-\mu}{\sigma}\right\}^2\right]\left[\int_0^x \lambda\exp\{-\lambda s\}\,ds\right]dx$$

$$= \int_{-\infty}^{\infty} \frac{1}{\sigma\sqrt{2\pi}}\exp\left[-\frac{1}{2}\left\{\frac{x-\mu}{\sigma}\right\}^2\right][1-\exp\{-\lambda x\}]\,dx$$

$$= 1 - \int_{-\infty}^{\infty} \frac{1}{\sigma\sqrt{2\pi}}\exp\left[-\frac{1}{2}\left\{\frac{x-\mu}{\sigma}\right\}^2\right]\exp\{-\lambda x\}\,dx.$$

This expression appears intractable at first sight, but the exponential terms can be combined and the exponent expressed in a way which allows a neat simplification to be made based on knowledge of the form of the normal distribution. The combined exponent is

$$-\frac{1}{2}\left\{\frac{(x-\mu)^2+2\sigma^2\lambda x}{\sigma^2}\right\} = -\frac{1}{2}\left\{\frac{(x-[\mu-\sigma^2\lambda])^2-\sigma^4\lambda^2+2\sigma^2\lambda\mu}{\sigma^2}\right\}$$

$$= -\frac{1}{2}\left\{\frac{(x-[\mu-\sigma^2\lambda])^2}{\sigma^2}\right\} + \frac{\{\sigma^2\lambda^2-2\lambda\mu\}}{2}.$$

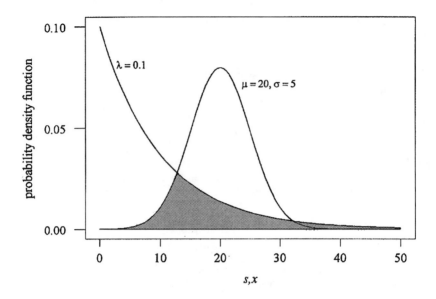

Figure 7.2. Interference region for Example 7.6

The first part of the exponent yields a normal distribution exponent and the second part can be taken out of the integral because it is independent of x. So

$$R = 1 - \exp\left[\frac{1}{2}\sigma^2\lambda^2 - \lambda\mu\right]\int_{-\infty}^{\infty}\frac{1}{\sqrt{2\pi}}\exp\left[-\frac{1}{2}\left\{\frac{x-[\mu-\sigma^2\lambda]}{\sigma}\right\}^2\right]dx$$

$$= 1 - \exp\left[\frac{1}{2}\sigma^2\lambda^2 - \lambda\mu\right].$$

It is only in a few such special cases that a tractable analytic form for R can be obtained.

7.8   Normally distributed stress and strength

Given stress S with mean μ_S and variance σ_S^2 and strength X with mean μ_X and variance σ_X^2,

$$f_S(s) = \frac{1}{\sigma_S\sqrt{2\pi}}\exp\left[-\frac{1}{2}\left(\frac{s-\mu_S}{\sigma_S}\right)^2\right], \quad -\infty < s < \infty$$

$$f_X(x) = \frac{1}{\sigma_X\sqrt{2\pi}}\exp\left[-\frac{1}{2}\left(\frac{x-\mu_X}{\sigma_X}\right)^2\right], \quad -\infty < x < \infty.$$

The interference random variable is $Y = X - S$ and is normally distributed with mean

$$\mu_Y = \mu_X - \mu_S$$

and standard deviation

$$\sigma_Y = \sqrt{\sigma_X^2 + \sigma_S^2}.$$

The reliability is given by

$$R = P(Y > 0) = \int_0^\infty \frac{1}{\sigma_Y\sqrt{2\pi}}\exp\left[-\frac{1}{2}\left(\frac{y-\mu_Y}{\sigma_Y}\right)^2\right]dy.$$

Standardizing y by letting $z = (y - \mu_Y)/\sigma_Y$ gives $\sigma_Y dz = dy$, and when $y = 0$, z is

$$z_0 = \frac{0 - \mu_Y}{\sigma_Y} = -\frac{\mu_X - \mu_S}{\sqrt{\sigma_X^2 + \sigma_S^2}} \tag{7.13}$$

Therefore

$$R = \int_{z_0}^\infty \frac{1}{\sqrt{2\pi}}e^{-z^2/2}dz = 1 - \Phi(z_0), \tag{7.14}$$

where Φ is the standard normal distribution function.

Reliability can be extremely sensitive to input errors. A high degree of accuracy is required for μ_Y and σ_Y in order to arrive at a good estimate of the reliability.

Example 7.7: Suppose a component has normally distributed strength with mean 400 MPa and standard deviation 50 MPa and is subject to stress which is also normally distributed but with mean 250 MPa and standard deviation 50 MPa. The value of z_0 is

$$- \frac{400 - 250}{\sqrt{50^2 + 50^2}} = -2.12.$$

So

$$R = 1 - \Phi(-2.12) = 1 - 0.017 = 0.983.$$

Example 7.8: In fact it is often better in practice to approach the problem from a different viewpoint. It is more likely that a certain value is perceived to be the minimum acceptable reliability, and then the question is about the strength parameters necesssary to meet this requirement. Say a component is required which will be subject to normally distributed stress, mean 200 MPa, standard deviation 30 MPa. Given a minimum reliability of 0.99, it can be deduced from normal tables that z_0 must be -2.327 at most. Substituting into (7.13), we have

$$2.327 = \frac{\mu_X - 200}{\sqrt{\sigma_X^2 + 30^2}}.$$

If either μ_X or σ_X is fixed then a limit for the other parameter is determined. There may, however, be some choice available and a component with mean strength lower than a competitor may be more advantageous if it also has lower standard deviation.

Example 7.9: A certain tension element has circular cross-section and, due to manufacturing tolerance, its diameter, D, is a random variable. The load, P, acting on the element is also a random variable. The ultimate tensile strength, X, of the material used for the element is a random variable because the properties of the material vary. Failure is known to occur by tensile fracture. For simplicity, stress and strength will be assumed to be normally distributed random variables. Experimental information yields:

$$\mu_P = 17790N, \sigma_P = 445N,$$

$$\mu_X = 689N/mm^2, \sigma_X = 34.5N/mm^2,$$

and the specified reliability is 0.9999.

Stress is given by $S = P/A$, where $A = \pi R^2$ and $R = D/2$. By (7.9), $\mu_A = E(A) \approx \pi[E(R)]^2 = \pi\mu_R^2$; and from (7.10), $\sigma_R^2 \approx (2\pi\mu_R)^2\sigma_A^2$.

Suppose the tolerance on the radius of the circular cross-section is such that $\sigma_R = \alpha\mu_R$, in other words the *coefficient of variation*, σ_R/μ_R, is α. Then

$$\mu_S \approx \mu_P/\mu_A \qquad \text{(from (7.9))}$$
$$\approx \mu_P/(\pi\mu_R^2)$$

and

$$\sigma_S^2 \approx \sigma_P^2(1/\mu_A)^2 + \sigma_A^2(-\mu_P/\mu_A^2)^2 \qquad \text{(using (7.10))}$$
$$= \frac{\pi^2\mu_R^4[\sigma_P^2 + 4\alpha^2\mu_P^2]}{\pi^4\mu_R^8}$$
$$= \frac{\sigma_P^2 + 4\alpha^2\mu_P^2}{\pi^2\mu_R^4}.$$

From normal distribution tables, for the specified reliability,

$$z_0 = -\frac{\mu_X - \dfrac{\mu_P}{\pi\mu_R^2}}{\sqrt{\sigma_X^2 + \sigma_S^2}} = -3.72.$$

If we let $\alpha = 0.005$, and substitute the known information, the equation simplifies to

$$\mu_R^4 - 17.03\mu_R^2 + 69.27 = 0$$

There are two positive roots; $\mu_R = 3.21$ mm, which gives the specified reliability, and $\mu_R = 2.59$ mm, which results in a reliability of 0.0001 (the complement of the specified reliability).

The distribution of stress and strength variables may also be functions of time. In particular it is likely that the average strength reduces over time and may be accompanied by increasing standard deviation. One way of modelling this time dependency is to apply a scaling factor to the hazard rate.

7.9   Safety factors and reliability

Safety factors were referred to in Section 7.5. A mean factor of safety is defined as

$$\eta = \frac{\mu_X}{\mu_S} \,,$$

and the coefficients of variation (see Section 7.8) of strength and stress, denoted γ_X and γ_S respectively, are

$$\gamma_X = \frac{\sigma_X}{\mu_X} \text{ and } \gamma_S = \frac{\sigma_S}{\mu_S}.$$

The coefficient of variation is equal to one for the exponential distribution. For most other distributions it is less than one.

It is straightforward to show that z_0 in (7.13) may be expressed as

$$z_0 = \frac{1 - \eta}{\sqrt{\gamma_X^2 \eta^2 + \gamma_S^2}}.$$

For a given η the reliability will decrease as σ_X and σ_S increase. Ideally a large η and low σ_X and σ_S are required. The production process controls σ_X and σ_S is largely determined by the operating environment. To increase the safety factor μ_X needs to be much larger than μ_S. *Overdesigning* of components will increase μ_X and *derating* reduces μ_S. The latter may be achieved, for example, by a system design which distributes the load over a number of parallel components.

The factor of safety may also be treated as a random variable $N = X/S$. If the distributions of X and S are known then the distribution of N may be found using the methods of Section 7.4, and is illustrated in Example 7.10.

Example 7.10: Let X_1 and X_2 be independent, identically distributed $N(0,1)$ random variables and $u_1(x_1, x_2) = x_1 + x_2 = y_1$, $u_2(x_1, x_2) = x_1/x_2 = y_2$. Of interest is u_2, but some form for a u_1 is required in order that the mathematics can yield the behaviour of u_2. The most mathematically convenient form for u_1 is chosen.

Rearranging these equations gives

$$x_1 = \frac{y_1 + y_2}{1 + y_2}, \quad x_2 = \frac{y_1}{1 + y_2}$$

and hence

$$|J(\mathbf{y})| = \frac{|y_1|}{(1 + y_2)^2}.$$

Since X_1 and X_1 are independent,

$$f\mathbf{X}(x_1, x_2) = \frac{1}{\sqrt{2\pi}} exp\left(-\frac{1}{2}x_1^2\right)\frac{1}{\sqrt{2\pi}} exp\left(-\frac{1}{2}x_2^2\right)$$

$$= \frac{1}{2\pi} exp\left[-\frac{1}{2}(x_1^2 + x_2^2)\right].$$

So,

$$f\mathbf{Y}(\mathbf{y}) = \frac{|y_1|}{(1 + y_2)^2}\frac{1}{2\pi} exp\left[-\frac{y_1^2(1 + y_2^2)}{2(1 + y_2)^2}\right].$$

The distribution of $Y_2 = X_1/X_2$ is found by integrating $f_Y(\mathbf{y})$ with respect to y_1 over $(-\infty, \infty)$. This yields $f_{Y_2}(y_2) = 1/[\pi(1 + y_2^2)]$, which is the *Cauchy distribution*.

7.10 Graphical approach for empirically determined distributions

This is a method of computing reliability where no basis exists for assuming any specific distribution for either stress or strength but there are experiments which have yielded sufficient empirical data.
Define

$$G = \int_s^\infty f_X(x)dx = 1 - F_X(s),$$

the strength reliability function, and

$$H = \int_0^s f_S(u)du = F_S(s),$$

the stress distribution function.

The range of H is from 0 to 1 and the probability element $\delta H \approx f_S(s)\delta s$. Substituting into Equation (7.11), the reliability is given by

$$R = \int_0^1 G dH.$$

The area under a plot of G against H would represent the reliability of the component. Estimates of $F_X(s)$ and $F_S(s)$ can be based on the strength and stress data and hence G and H. Plotting these values of G and H against each other and measuring the area graphically determines the reliability.

Example 7.11: The experimental data in Table 7.1 are given by Kapur and Lamberson (1977). In a stress analysis of a component 10 observations were made under simulated operating conditions, and a strength analysis yielded 14 values for strength. Smooth curves (Figures 7.3 and 7.4) through the $\hat{F}$ values yielded by the data are used as estimators of the unknown distribution functions. Values for $\hat{F}$ have been calculated using $(i-0.3)/(n+0.4)$ (see Section 3.2). Estimated values of G and H may then be computed. The area shown under the plot in Figure 7.5 represents the estimated reliability of the component. Measuring this using a 'trapezium rule' yields the slightly conservative estimate of 0.913.

Example 7.12: To verify this approach, consider the case of Example 7.6, where component strength is normally distributed and applied stress exponentially distributed. It has been shown that the reliability of this component is given by

$$R = 1 - \exp\left[\frac{1}{2}\sigma^2\lambda^2 - \lambda\mu\right],$$

where μ and σ are the normal distribution parameters and λ the exponential parameter. Suppose $\mu = 20$, $\sigma = 5$ and $\lambda = 0.1$. The exact reliability is $1 - \exp[-1.875] = 0.8466$. Figure 7.6 shows a plot of G against H

X	$\hat{F}_X(x)$	S	$\hat{F}_S(s)$	X, S	$G = 1 - \hat{F}_X$	$H = \hat{F}_S$
33.8	0.049	20.75	0.067	0	1.00	0.00
34.3	0.118	23.60	0.163	10	1.00	0.00
35.4	0.188	24.50	0.260	15	1.00	0.00
35.9	0.257	26.25	0.356	20	1.00	0.05
36.0	0.326	26.50	0.452	25	1.00	0.28
36.0	0.396	27.50	0.548	30	1.00	0.74
36.8	0.465	29.25	0.644	32	0.99	0.78
37.0	0.535	30.00	0.740	33	0.96	0.82
37.1	0.604	33.75	0.837	34	0.91	0.85
37.3	0.674	37.50	0.933	35	0.81	0.88
38.2	0.743			36	0.60	0.90
38.5	0.813			37	0.44	0.92
40.0	0.882			38	0.25	0.94
42.0	0.951			39	0.16	0.96
				40	0.12	0.98
				41	0.08	0.99
				42	0.05	1.00

Table 7.1. G and H calculations for Example 7.11

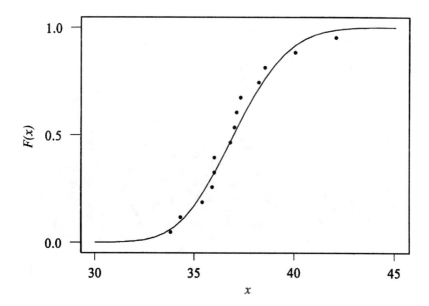

Figure 7.3. Estimated strength distribution for Example 7.11

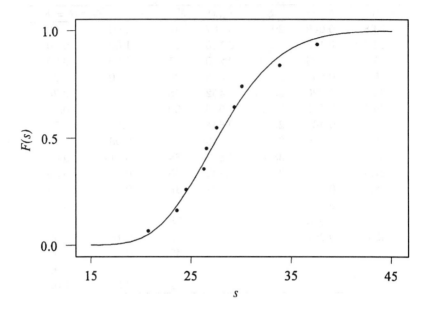

Figure 7.4. Estimated stress distribution for Example 7.11

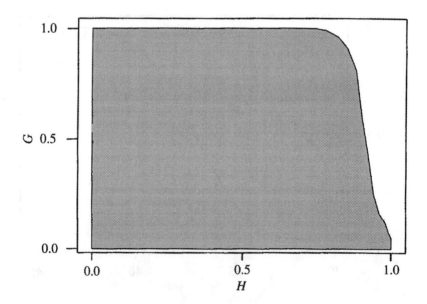

Figure 7.5. Plot of G against H for Example 7.11

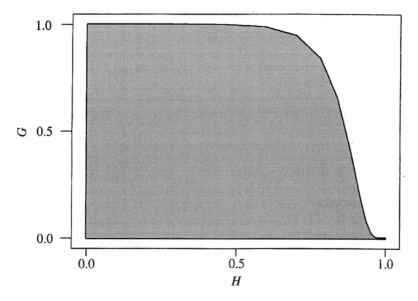

Figure 7.6. Plot of G against H for Example 7.12

using 16 stress and strength distribution function values calculated with the given parameters. The reliability estimated from the shaded area of the plot is 0.8444.

CHAPTER 8

Maintenance Strategies

8.1   Maintained systems

Maintenance is any form of intervention in a system with the intent either to keep the system in operational state or to restore it to an operational state after failure. Most systems require *preventive* and *corrective* maintenance; the latter occurs after failure, while the former is designed to prevent failure and so lengthen the MTTF. When a system has some measure of redundancy, unit failure may not necessarily result in system failure, and therefore repairing failed units is partly corrective, partly preventive.

For maintained systems, questions of interest include:

- What is the mean time to first system failure?
- What is the long-term availability?
- How often does the system go 'down' in a given interval of time?

Clearly the repair time distribution and the costs of failure are very important. The maintenance of a system may involve repair only after the system has failed, or some repair may be possible in systems which have redundancy and which may therefore be able to continue to function while some components are repaired.

Preventive maintenance may include scheduled inspections or condition-based maintenance. The latter is concerned with signs of future failure and identification of any trend in modes of failure.

> **Example 8.1:** Consider a two-unit passive redundant system with a single repair operative. System failure only occurs if a unit is required while it is still being repaired, that is, if the lifetime, T, of the unit in operation is less than the time, X, required to repair the other unit.

Let $f(t)$ be the unit lifetime distribution and $g(x)$ the distribution of repair time. Then

$$P(X > T) = \int_0^\infty f(t) \left[\int_t^\infty g(x) \mathrm{d}x \right] \mathrm{d}t .$$

There is a direct analogy here with the stress–strength model of (7.11). Given that X is greater than T, the expected length of time until the repair is completed is the residual mean lifetime

$$m_X(t) = \frac{\int_t^\infty R_X(x) \mathrm{d}x}{R_X(t)}$$

(this expression reduces to (1.8) when $t = 0$). To take the case of exponential lifetimes, rate λ, and exponential repair times, rate μ,

$$P(X > T) = \frac{\lambda}{\lambda + \mu} = \frac{\text{MTTR}}{\text{MTTR} + \text{MTTF}},$$

where MTTR is mean time to repair, and

$$m_X(t) = \frac{1}{\mu} = \text{MTTR} .$$

The expected number of failures in operating time t is $t\lambda$ and a proportion $\lambda/(\lambda + \mu)$ of these will result in down time.

The expected total down time is

$$t\lambda \left(\frac{\lambda}{\lambda + \mu} \right) \frac{1}{\mu} .$$

Clearly it is desirable to have MTTR considerably shorter than MTTF, that is, μ much larger than λ.

If failed parts in a redundant system are repaired then a considerable increase in reliability can be achieved since the system is only vulnerable during repair time. In the case of exponential repair times with $\mu = 1/\text{MTTR}$ a two-unit system gives

$$\text{MTTF (active)} = \frac{3}{2\lambda} + \frac{\mu}{2\lambda^2}$$

$$\text{MTTF (standby)} = \frac{2}{\lambda} + \frac{\mu}{\lambda^2}.$$

With μ many times greater than λ, repair would have a significant effect on the reliability of the system.

8.2   Availability

The *availability function*, $A(t)$, represents the probability that a system is operational at time t. For non-repairable systems availability is the same as reliability, that is, $A(t)$ is the same as the reliability function, $R(t)$. When a system is repairable, it will alternate over time between being 'up' and 'down'. Whereas the time scale for the reliability function is a period of continuous 'up' time, availability is measured over a period potentially containing 'down' time.

'Mission availability' is the average availability in the short term:

$$A(t, T) = \frac{1}{T} \int\limits_{t}^{t+T} A(t)\mathrm{d}t.$$

The 'long-term' or 'steady-state' availability is the limit of $A(t)$ as $t \to \infty$.

$$\lim_{t \to \infty} \frac{1}{t}\int\limits_{0}^{t} A(t)\mathrm{d}t,$$

which is effectively the limit of $A(t)$ as $t \to \infty$

This will be referred to as A_∞.

Steady-state availability is also given by MTBF/(MTBF + MTTR), where MTBF is the *mean time between failures* (see Section 8.4) but it should be noted that this value is only approached after a large number of repairs. Note also that availability does not supply information about the *number of repairs*.

For a series system of n units each with exponential lifetime, parameter λ, and repair time distribution exponential with rate μ, the availability is

$$A(t) = \frac{\mu}{n\lambda + \mu} + \frac{n\lambda}{n\lambda + \mu} e^{-(n\lambda + \mu)t}. \tag{8.1}$$

Figure 8.1 illustrates (8.1) for $n = 2$, $\lambda = 1$, $\mu = 10$.

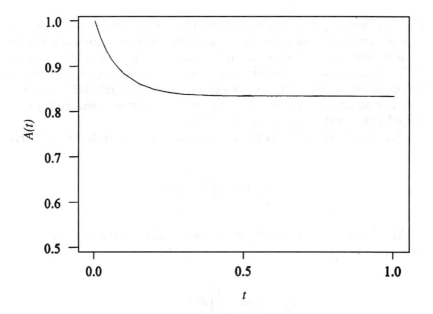

Figure 8.1. Availability function for a two-unit series system

The long-term availability, A_∞, in this case is

$$A_\infty = \frac{\mu}{n\lambda + \mu}.$$

This is also given by

$$\frac{\text{MTBF}}{\text{MTBF} + \text{MTTR}} = \frac{1/n\lambda}{1/n\lambda + 1/\mu} = \frac{\mu}{n\lambda + \mu}.$$

For the values of n, λ and μ used in Figure 8.1, A_∞ is approached fairly rapidly.

8.3   Markovian systems

Where it is possible to repair the system to former operational conditions, it becomes possible to revisit all system states; there is no longer a 'final', or *absorbing*, state. The Markov diagram then depicts continuous movement of the system.

Example 8.2: Consider the two-unit system of Example 8.1 with exponential unit life and repair times, rates λ and μ, respectively (Figure 8.2). The system states are: S_0, 2 units available; S_1, 1 unit available; S_2, 0 units available. Following the methodology of Section 6.7, the differential equations describing transitions over time are

$$\begin{bmatrix} P_0'(t) \\ P_1'(t) \\ P_2'(t) \end{bmatrix} = \begin{bmatrix} -\lambda & \mu & 0 \\ \lambda & -\mu-\lambda & \mu \\ 0 & \lambda & -\mu \end{bmatrix} \begin{bmatrix} P_0'(t) \\ P_1'(t) \\ P_2'(t) \end{bmatrix}. \qquad (8.2)$$

Taking Laplace transforms, and assuming the system is in state S_0 at $t = 0$,

$$\begin{bmatrix} s+\lambda & -\mu & 0 \\ -\lambda & s+\mu+\lambda & -\mu \\ 0 & -\lambda & s+\mu \end{bmatrix} \begin{bmatrix} P_0^*(s) \\ P_1^*(s) \\ P_2^*(s) \end{bmatrix} = \begin{bmatrix} 1 \\ 0 \\ 0 \end{bmatrix}.$$

A general expression for the equations to be solved is

$$\mathbf{M}\pi = \mathbf{d},$$

where $\mathbf{M}$ is an $n \times n$ matrix which is a function of s and the model parameters, n being the number of non-absorbing system states, π is

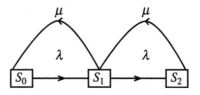

Figure 8.2. Markov diagram for Example 8.2

a vector of the $P_i^*(s)$ and **d** is the probability vector at time $t = 0$. All entries in **d** are zero except for the probability of being in the initial state, which is 1. The required solution is

$$\pi = \mathbf{M}^{-1}\mathbf{d}. \tag{8.3}$$

This exercise may well be non-trivial. The equations may be solved by substitution but the matrix inversion is now possible quite neatly by use of computer algebra packages such as DERIVE, MATHEMATICA or MATLAB. There are, however, even with these tools, limitations on the size of n. We in fact only need the rth column of $\mathbf{M}^{-1}$, where r is the row of **d** where the '1' is positioned. This yields, for Example 8.2,

$$P_0^*(s) = \frac{(s+\mu)^2 + \lambda s}{\phi}, \; P_1^*(s) = \frac{(s+\mu)\lambda}{\phi}, \; P_2^*(s) = \frac{\lambda^2}{\phi},$$

where $\phi = s[(s+\mu)^2 + (2s+\mu)\lambda + \lambda^2] = s[s^2 + 2s(\mu+\lambda) + (\mu^2 + \lambda\mu + \lambda^2)]$. To invert the $P_i^*(s)$ it is necessary to expand the expression $1/\phi$ using partial fractions (see the Appendix). Here we can write $1/\phi$ as

$$\frac{1}{s(s+a)(s+b)} = \frac{1}{abs} + \frac{1}{a(a-b)(s+a)} - \frac{1}{b(a-b)(s+b)},$$

where $ab = \mu^2 + \lambda\mu + \lambda^2$ and $a + b = 2(\mu + \lambda)$. Now inverting $P_2^*(s)$ yields

$$P_2(t) = \lambda^2 \left[\frac{1}{ab} + \frac{1}{a-b} \left\{ \frac{e^{-at}}{a} - \frac{e^{-bt}}{b} \right\} \right].$$

Inverting $P_0^*(s)$ and $P_1^*(s)$ requires terms involving s to be eliminated from the numerator. For example,

$$P_1^*(s) = \frac{s\lambda}{\phi} + \frac{\mu\lambda}{\phi} = \frac{\lambda}{(s+a)(s+b)} + \frac{\mu\lambda}{s(s+a)(s+b)} \, .$$

However $P_2^*(t)$ is the most useful term here as it represents the *unavailability*.

Example 8.2 (continued): Availability is to do with the system being in a functioning state, so in this case

$$A(t) = P_0(t) + P_1(t) = 1 - P_2(t)$$

$$= 1 - \frac{\lambda^2}{\mu^2 + \lambda\mu + \lambda^2} - \frac{\lambda^2}{(a-b)}\left[\frac{e^{-at}}{a} - \frac{e^{-bt}}{b}\right].$$

Letting t tend to infinity yields

$$A_\infty = 1 - \frac{\lambda^2}{\mu^2 + \lambda\mu + \lambda^2} = \frac{\mu^2 + \lambda\mu}{\mu^2 + \lambda\mu + \lambda^2} \, .$$

In the state-space analysis of systems we can alternatively find A_∞ by using the fact that all $P_i'(t)$ are zero in the steady state. This leads to direct relationships between the steady state P_i, and given that $\sum_i P_i = 1$, each P_i can be evaluated. A_∞ is given by the sum of those P_i relating to operational states.

Example 8.2 (continued): The steady-state situation is given by letting $P_i'(t) = 0$ in (8.2). Denoting the steady-state probabilities P_i, we have

$$\begin{bmatrix} -\lambda & \mu & 0 \\ \lambda & -\mu-\lambda & \mu \\ 0 & \lambda & \mu \end{bmatrix}\begin{bmatrix} P_0 \\ P_1 \\ P_2 \end{bmatrix} = 0 \, .$$

These equations are linearly dependent and do not therefore have a unique solution. However, by definition $P_0 + P_1 + P_2 = 1$ and this equation is substituted for one of the above. So we have

$$\frac{\mu P_1}{\lambda} + P_1 + \frac{\lambda P_1}{\mu} = 1$$

from which P_1 can be determined and hence, by substitution, P_0 and P_2.

$$P_0 = \frac{\mu^2}{\mu^2 + \lambda\mu + \lambda^2}, \; P_1 = \frac{\lambda\mu}{\mu^2 + \lambda\mu + \lambda^2}, \; P_2 = \frac{\lambda^2}{\mu^2 + \lambda\mu + \lambda^2}.$$

So

$$A_\infty = P_0 + P_1 = \frac{\mu^2 + \lambda\mu}{\mu^2 + \lambda\mu + \lambda^2}. \tag{8.4}$$

The long-run availability is only achieved after many repairs and may for some systems be too far into the future to be a useful measure. Something similar to A_∞ may be achieved in the shorter term in a situation where there are a large number of similar systems, probably not all commissioned at the same time.

It can be useful to express A_∞ in terms of the unit *failure-to-repair* ratio, $\beta = \lambda/\mu$, where β is assumed to be small. For the n-unit series system $A_\infty = 1/(1 + n\beta)$. Using this terminology it is informative to consider the effect of the ability to carry out more than one repair simultaneously. Figure 8.3 shows the resulting Markov diagrams and long-term availabilities for two-unit redundant systems. In the desirable case of β small, the effect of more than one repair facility is very small and in that case unlikely to be cost-effective.

When there is more than one repair facility, there may be questions about how most efficiently to deploy those facilities in relation to:

(a)   more than one repairer working on a single failed unit;
(b)   repair facilities being transferable from one unit to another.

(a) is not always practically possible, and when it is, will not necessarily result in maximum utilization of resource. In many cases it is likely that, if each repairer repairs at rate μ, two repairers working together on the same unit will repair at rate $\mu(1 + \alpha)$, where $0 < \alpha < 1$, but some repair jobs, awkward for one repairman, might result in $\alpha > 1$. (b) is about whether it is better to have a one-to-one relationship between unit and repairer or to have repairers who may be assigned to any unit.

8.4   Mean time between failures (MTBF)

The MTBF is to do with the distribution of time between points when a system is restored from a failed to a working state. This may not be the same distribution as that for time to first system failure (recall

Two-unit passive redundant system

(a) with one repair facility:

$$A_\infty = \frac{1 + \beta}{1 + \beta + \beta^2}$$

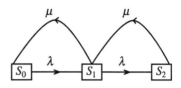

(b) with two repair facilities:

$$A_\infty = \frac{1 + \beta}{1 + \beta + \frac{1}{2}\beta^2}$$

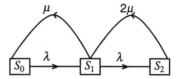

Two-unit active redundant system

(a) with one repair facility:

$$A_\infty = \frac{1 + 2\beta}{1 + 2\beta + 2\beta^2}$$

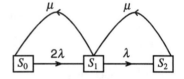

(b) with two repair facilities:

$$A_\infty = \frac{1 + 2\beta}{1 + 2\beta + \beta^2}$$

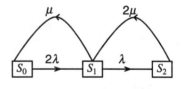

Figure 8.3. Markov diagrams for two-unit redundant systems

the modified renewal process of Section 5.6). When a system is restored at failure to the condition at $t = 0$, then the MTBF is the same as MTTFF (mean time to *first* system failure). This would be true for a system with no redundancy, such as a pure series system. Where there is redundancy in a system and repair at component level is possible, system operation can be restored while further repair is in progress. In Example 8.2 the system starts in state S_0, and will revisit S_0 at various times in the future but typical system operation is to do with transitions between states S_1 and S_2, S_2 being the state which would be absorbing if there were no repair at system level and S_1 the state which results immediately after system repair. To find the MTBF, we consider the reliability function for the time to system failure taking S_1 as the initial state and omitting 'system' repair, that is, the transition from S_2 to S_1.

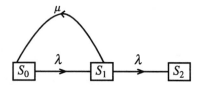

Figure 8.4. Markov diagram for a two-unit passive redundant system with repair at component level only

The Markov diagram is shown in Figure 8.4 and the initial probability vector $\mathbf{d}^{\mathrm{T}} = [0 \quad 1 \quad 0]$. Equations (8.2) reduce to

$$\begin{bmatrix} P_0'(t) \\ P_1'(t) \end{bmatrix} = \begin{bmatrix} -\lambda & \mu \\ \lambda & -\mu-\lambda \end{bmatrix} \begin{bmatrix} P_0'(t) \\ P_1'(t) \end{bmatrix}$$

and (8.3) becomes

$$\begin{bmatrix} P_0^*(s) \\ P_1^*(s) \end{bmatrix} = \begin{bmatrix} s+\lambda & -\mu \\ -\lambda & s+\mu+\lambda \end{bmatrix}^{-1} \begin{bmatrix} 0 \\ 1 \end{bmatrix}$$

This gives

$$P_0^*(s) = \frac{\mu}{(s+\lambda)(s+\lambda+\mu)-\lambda\mu}, \quad P_1^*(s) = \frac{s+\lambda}{(s+\lambda)(s+\lambda+\mu)-\lambda\mu}.$$

The reliability function for the time between failures has transform

$$R_b^*(s) = P_0^*(s) + P_1^*(s) = \frac{s+\lambda+\mu}{s^2 + (2\lambda+\mu)s + \lambda^2}. \tag{8.5}$$

The MTBF is given by setting $s = 0$, (see (6.5)), hence

$$\text{MTBF} = \frac{\lambda + \mu}{\lambda^2} = \frac{1}{\lambda} + \frac{\mu}{\lambda^2}.$$

Now in Section 8.1 the mean time to first failure for this two-unit passive system was given as

$$\frac{2}{\lambda} + \frac{\mu}{\lambda^2}.$$

This is the MTBF with, in addition, the expected time to failure for unit 1. In 'new' condition the system takes a little longer to its first failure than it does on average between subsequent failures. The smaller λ is relative to μ, the more insignificant is the difference between the two mean failure times. There are, however, some systems for which the time spent 'out of action' is quite substantially larger than that in operation and therefore differences between MTTFF and MTBF may need to be very specifically catered for.

Returning to (8.5), if we wished to look at the distribution of time between failures, we need to find $R_b(t)$. The inversion of $R_b^*(s)$ again requires first removing s from the numerator of the expression by decomposing into partial fractions.

Example 8.3: To illustrate the above in relatively simple terms we will let $\lambda = 1$ and $\mu = 5$. Now

$$R_b^*(s) = \frac{s+6}{s^2 + 7s + 1} = \frac{s+6}{(s+6.85)(s+0.15)},$$

to reasonable accuracy. We now write the numerator as $s + 6.85 - 0.85$ and this allows the following separation.

$$R_b^*(s) = \frac{1}{(s+0.15)} - \frac{0.85}{(s+6.85)(s+0.15)}.$$

The inversion now follows using Table 5.2:

$$R_b^*(t) = e^{-0.15t} - \frac{0.85}{6.85 - 0.15}(e^{-0.15t} - e^{-6.85t})$$

$$= 0.8731\ e^{-0.15t} + 0.1269\ e^{-6.85t}.$$

Figure 8.5 shows $R_b(t)$. With $\lambda = 1$, the time scale is effectively in multiples of the mean failure time of a single unit. The effect of reducing the failure-to-repair ratio, β, that is increasing the value of μ, is to lengthen the time between failures.

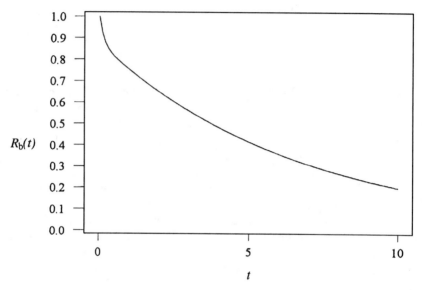

Figure 8.5. Reliability function for the time between failures for a two-unit passive redundant system

8.5 Age replacement

A common type of preventive maintenance involves anticipating the likely occurrence of failure of a component and replacing it before failure and the possible penalties resulting from failure. There is, however, only sense in doing this if the replacement component has a lower hazard rate than the one it replaces. For example, there is no point in planned replacement of single components or systems with exponential lifetimes because the hazard function is constant. No matter how old the component, the tendency to fail is unchanged. If the new component has exponential lifetime but with lower hazard, then a case for replacement might be made. Rewriting software, which commonly has a lifetime with decreasing hazard function, may well result in higher, not fewer, numbers of failures in the short term.

Both replacement and failure will have costs attached to them. These may frequently be expressed in financial terms, but equally some costs might be better expressed in terms of time. If an age-replacement policy is to be implemented, clearly of interest is the 'best age' to adopt, which in practice is determined by some optimum cost condition. It is of course assumed that the cost of a replacement is less than the cost of a failure.

Suppose a component has lifetime with reliability function $R(t)$ and is replaced either at failure or on reaching age τ, whichever occurs first. The cost of a *planned* replacement is c_1 and the cost of replacement at failure $c_2 (> c_1)$. If Y is the cost of *any* replacement, then the expected value of Y is

$$E(Y) = c_1 R(\tau) + c_2[1 - R(\tau)]. \qquad (8.6)$$

There is also the question of how often a replacement is made. The cost over a given period is the product of cost and the number of replacements, or, to put this another way, the cost per unit time is of interest and what we seek to minimize.

Let X be the length of time between replacements. For component lifetimes t less than τ, $X = t$; for those greater than τ, $X = \tau$. So

$$E(X) = \int_0^\infty x f(x) dx = \int_0^\tau x f(x) dx + \int_\tau^\infty \tau f(x) dx,$$

where $f(x)$ is the probability density function of the component lifetime distribution, that is, $d/dx\,[1 - R(x)]$. Integrating by parts and expressing in terms of the function R, we have

$$E(X) = \tau[1 - R(\tau)] - \int_0^\tau [1 - R(x)] dx + \tau R(\tau)$$

$$\qquad (8.7)$$

$$= \int_0^\tau R(x) dx.$$

With no age replacement, that is $\tau = \infty$, $E(X)$ is the mean time to fail as in (1.8).

Over a long period of time, a simple cost measure can be given by $g(\tau; c_1, c_2) = E(Y)/E(X)$, that is, (8.6) divided by (8.7). The optimization is with respect to τ conditional on the values of c_1 and c_2. In the short term, the analysis is more complicated but practically of less interest.

Example 8.4: Consider a component with lifetime having reliability function $R(t) = e^{-t^2/2}$ (Weibull, $\alpha = \sqrt{2}, \beta = 2$).

$$g(\tau;c_1, c_2) = \frac{c_1 R(\tau) + c_2[1 - R(\tau)]}{\int_0^\tau R(x)dx}$$

$$= \frac{c_2 - (c_2 - c_1)R(\tau)}{\int_0^\tau R(x)dx} \qquad (8.8)$$

$$= \frac{c_2 - (c_2 - c_1)e^{-\tau^2/2}}{\sqrt{2\pi}[\Phi(\tau) - 0.5]}$$

where $\Phi(.)$ is the standard normal distribution function (this is simply a handy feature of this example).

Now suppose that the cost of failure is twice the cost of planned replacement. This can be simply represented by letting $c_1 = 1$ and $c_2 = 2$. We can think of c_1 as some standard cost. The function $g(\tau)$ is not in analytic form so a numerical optimum is obtained and illustrated graphically in Figure 8.6. It can be seen that the best replacement time is at approximately $\tau = 1.5$. The optimum cost measure is 1.543. If there were no planned replacement, that is if $\tau = \infty$ in (8.8), then the cost measure would be $4/\sqrt{2\pi} \approx 1.596$. Age replacement reduces costs by 3.3%. If $c_2 = 3$ then the optimum τ is 1.0 (note that the mean time to fail for these components is 1.25) with cost measure 2.09, representing a 12.7% reduction in costs. It is interesting to see from (8.6) the proportion of replacements which are planned, namely $R(\tau)$, which is dependent on τ only, not the costs. For $\tau = 1$, 60% are age replacements.

Example 8.5: From Example 2.3 a passive redundant system with two constant failure rate units has $R(t) = e^{-\lambda t}(1 + \lambda t)$. Without loss of generality, we will let $\lambda = 1$ and $c_1 = 1$ with $c_2 = k$, k representing the order of the difference between replacement and failure costs. Equation (8.8) becomes

$$g(\tau) = \frac{k - (k - 1)e^{-\tau}(1 + \tau)}{2 - 2e^{-\tau} + \tau e^{-\tau}}.$$

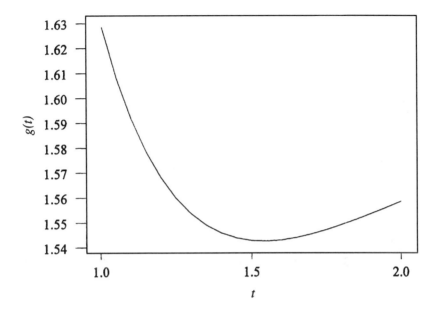

Figure 8.6. Cost function for Example 8.4

Figure 8.7 shows how the optimum value of τ varies with the value of k. It should be emphasized that the replacement here is at system level. The components have exponential lifetimes so age replacement at component level is not worthwhile, but the system has increasing hazard so replacing the system is worthwhile. For example, overhead projectors are usually equipped with a spare bulb. The 'life' of the projector can be thought of in terms of a two-unit passive redundant system. If the bulbs have exponential lifetimes an age replacement policy could be applied on a pair-of-bulbs basis. This would mean that inspection ensured that there were two working bulbs. There would be no point in replacing a working bulb because of its constant hazard rate, but any failed bulb would be replaced.

The relative merits of different maintenance strategies will depend not only on the timing and unit costs of procedures such as age replacement, but also on the lifetime distribution of components. Most practical systems would also involve several components which might fail.

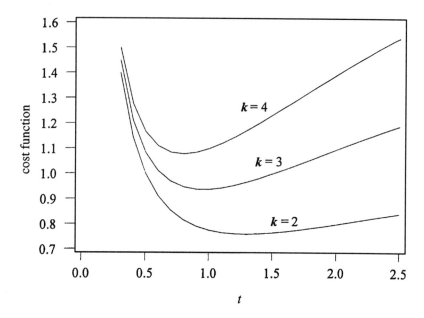

Figure 8.7. Cost functions for k = 2, 3, 4 in Example 8.5

Example 8.6: Suppose a new type of bus shelter is to be installed in a metropolitan area, and this shelter has a light which is permanently on. The information supplied by the manufacturer is only that 95% of bulbs have a lifetime exceeding 100 hours, that is, for lifetime T, $R(100) = 0.95$. Nothing else is known about the lifetime distribution of the bulbs. A decision to apply age replacement to the bulbs would need to be based on an assumption of increasing hazard. Given that to be a reasonable assumption and shelters are to be inspected every k hours, some in-service reliability information is required together with a lifetime model in order to obtain some sort of optimal k.

Say, arbitrarily, that the shelters are inspected after 1000 hours and 30% of bulbs found to have failed. It is likely that a better performance rate is required, say 20% failure at inspection, but determining k in that case depends on a model assumption for the lifetimes. Two possibilities follow:

(a) lognormal: here we can say that log T has a normal distribution, say with parameters μ and σ. Given that $R(100) = 0.95$ and $R(1000) = 0.70$, μ and σ can be estimated. From standard normal tables, the standardized scores, z, corresponding to $\Phi(z) = 0.05$ and $\Phi(z) = 0.3$ are −1.645 and −0.5244. So,

$$-1.645 = \frac{\log 100 - \mu}{\sigma} ,$$

$$-0.5244 = \frac{\log 1000 - \mu}{\sigma} ,$$

yielding $\mu = 7.975$ and $\sigma = 2.05$. It follows that the standardized score for $\log k$ is

$$-0.8416 = \frac{\log k - 7.975}{2.05}$$

and solving yields $k = 518$ hours.

(b) Weibull: again using the recorded reliabilities, we have

$$0.95 = \exp[-(100/\alpha)^{\beta}],$$
$$0.70 = \exp[-(1000/\alpha)^{\beta}],$$

and hence $\alpha = 3433$ and $\beta = 0.84$. Now we could substitute these values into $R(k) = 0.80$ to give $k = 576$ hours, but it is worth noting that the value of β obtained is less than 2 and therefore indicates that the lifetime has decreasing hazard, so that age replacement would be nonsensical. More likely is the conclusion that a Weibull model is simply not appropriate.

Example 8.7: A floodlighting system has eight lamps. The bulbs for these lamps have normally distributed lifetimes, mean 1000 hours and standard deviation 150 hours. While it is ideal for all lamps to be functioning, there may be a tolerable level of performance even when some bulbs have failed. Consider the following two maintenance strategies:

(A) the fitter is called each time a bulb fails and the bulb is renewed;

(B) the fitter calls every 800 hours of operation and replaces all bulbs, failed or otherwise.

The standard charge each time the fitter visits is £50 under A and £30 under B. A new bulb costs £10. In purely monetary terms, a comparison of these strategies would be based on the average cost per unit system running time.

Under Strategy A, 8 bulbs would on average fail every 1000 hours and, given that the probability of more than one bulb failing at a time is negligible, each failure will cost £60 and hence the cost per running hour is $£(8 \times 60/1000) = £0.48$. Under B, the cost of each visit is £30 + 8 × 10 = 110. So the cost per running hour is £110/800 = £0.14.

Strategy B costs a lot less than A, but will result in a system which is not always fault-free. For what proportion of inspections will there be a fault-free system? This is given by the probability that all eight bulbs last 800 hours. All bulbs are new at the start of each interval so the same probability applies to each bulb. The probability of failure within 800 hours of life is given by $\Phi((800 - 1000)/150) = 0.0918$, where Φ is the standard normal distribution function. All eight bulbs last 800 hours with probability $(1 - 0.0918)^8 = 0.5048$. So almost 50% of inspections will involve replacement of failed bulbs.

Part of the equation here is what other 'costs', for example customer dissatisfaction, are involved in not having a fault-free system. Factors which will increase the proportion of fault-free time under age replacement include reducing the interval between visits and using a bulb with lifetime which has smaller variance. The latter is a more important factor than the value of the mean lifetime. To give an extreme example which illustrates this point, suppose the bulbs had mean 1000 hours and zero variance, then given all lights originally had new bulbs together, visits would take place at intervals just less than 1000 hours. The value of the mean contributes to the cost of maintenance but is not instrumental in obtaining a fault-free system.

Chapter 7 of Ascher and Feingold (1984) considers cost models for repairable systems. A mathematical treatment of maintenance policies may be found in Barlow and Proschan (1996), which is a replica of their 1965 work.

8.6   Scheduled maintenance

Scheduled maintenance is a preventive strategy and embraces age replacement, as in strategy B of Example 8.7. Here we will look at this practice from the point of view of the effect on the reliability function of the system.

Suppose that regular maintenance actions are taken at time points $t = i\tau$. Assume that the system is 'as new' after maintenance. Let $R(t)$

be the reliability of the system from 'as new' condition and $R_s(t)$ the reliability of the system under scheduled maintenance.

The reliability of the system is the probability of surviving without failure. The probability that the system is still working at $t = n\tau$, under the condition that the system is restored to 'as new' at each $i\tau$, is the product of the probabilities of surviving each interval τ, that is, $[R(\tau)]^n$:

time interval	reliability $R_s(t)$
$0 \to \tau$	$R(t)$
$\tau \to 2\tau$	$R(\tau)\, R(t - \tau)$
$2\tau \to 3\tau$	$R(\tau)\, R(\tau)\, R(t - 2\tau)$
.	
.	
$n\tau \to (n + 1)\tau$	$R(\tau)^n\, R(t - n\tau)$

Figure 8.8 shows the general effect of scheduled maintenance. Without maintenance the decline in the reliability is unchecked. Maintenance is putting off the moment when the system does actually fail.

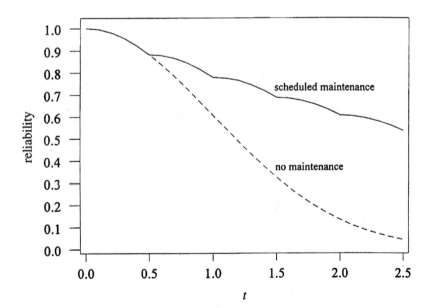

Figure 8.8. Reliability function under scheduled maintenance

The mean time to first failure is

$$
\text{MTTFF} = \int_0^\infty R_s(t)\,dt
$$

$$
= \left\{ \int_0^\tau R(t) + \int_\tau^{2\tau} R(\tau)R(t-\tau) + \int_{2\tau}^{3\tau} [R(\tau)]^2 R(t-2\tau) + \ldots \right\} dt
$$

$$
= \left\{ \int_0^\tau R(t) + R(\tau)\int_0^\tau R(t) + [R(\tau)]^2 \int_0^\tau R(t) + \ldots \right\} dt
$$

$$
= \sum_{n=0}^\infty R(\tau)^n \int_0^\tau R(t)\,dt
$$

$$
= \frac{\displaystyle\int_0^\tau R(t)\,dt}{1 - R(\tau)}
$$

(using the formula for the sum of an infinite geometric progression, since $R(\tau) < 1$ – see the Appendix).

Once again if our 'system' is a single unit with exponential lifetime, rate parameter λ, there is no gain in scheduled maintenance as $R_s(t)$ is always $e^{-\lambda t}$ and MTTFF is always $1/\lambda$.

Example 8.8: Suppose we have a two-unit active parallel system with $R(t) = 2e^{-\lambda t} - e^{-2\lambda t}$. Then

$$
\text{MTTFF} = \left[\frac{2e^{-\lambda t}}{-\lambda} - \frac{e^{-2\lambda r}}{-2\lambda} \right]_0^\tau \Big/ (1 - 2e^{-\lambda\tau} + e^{-2\lambda\tau})
$$

$$
= \frac{1}{\lambda}\left[-2e^{-\lambda\tau} + \frac{e^{-2\lambda\tau}}{2} + 2 - \frac{1}{2} \right] \Big/ (1 - 2e^{-\lambda\tau} + e^{-2\lambda\tau})
$$

$$
= \frac{1}{2\lambda}\frac{(3 - 4e^{-\lambda\tau} + e^{-2\lambda\tau})}{(1 - e^{-\lambda\tau})(1 - e^{-\lambda\tau})} = \frac{1}{2\lambda}\frac{(3 - e^{-\lambda\tau})}{1 - e^{-\lambda\tau}}.
$$

Note that as τ tends to infinity, that is, there is no scheduled maintenance, the MTTFF tends to $3/2\lambda$, the result following from Section 6.5. Scheduled maintenance at $\tau = 1/(2\lambda)$, half the mean lifetime of each unit, results in almost doubling the MTTFF.

The 'as new' assumption may be unrealistic: $R(\tau)$ may need adjusting at each τ interval. At each scheduled maintenance $R(t)$ is restored to 1 but may thereafter decline at a greater rate than previously; in other words, the hazard rate may be more markedly increasing at each τ interval. In general,

$$R_s(t) = R_1(\tau)\, R_2(\tau) \,\ldots\, R_{n+1}\, (t - n\tau).$$

Scheduled maintenance remains worthwhile all the time $R_s(t)$ remains greater than $R(t)$, as in Figure 8.8. Improving the $R_i(t)$ or reducing the value of τ are ways of making the policy more effective.

8.7   Systems with failure detection/fail-safe devices

While failure of a system may not ultimately be preventable, it can be possible to limit the effects of failure by advance warning or *fail-safe* mechanisms. However, such devices are themselves subject to failure and therefore the same issues concerning reliability apply. Usually the motivation for using these devices is that the system is in a safety-critical area.

The following example is a simplification of a Markovian system described in Soper and Wolstenholme (1993).

Example 8.9: Figure 8.9 shows a system comprising three units which run concurrently. Failure of the main unit, M, results in a dangerous condition. Unit D provides early warning of failure and in that situation unit F overrides unit M with a safe output. System performance depends critically on D and F. Unless these units operate efficiently, there is no improvement (in terms of failing safely) over M on its own.

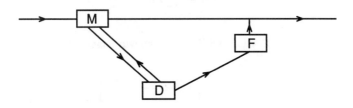

Figure 8.9. System illustration for Example 8.9

All units are considered to have exponential times to failure. Let M fail at rate λ and D and F fail at rate $k\lambda$, where k is considerably less than 1. System states are defined as follows:

	M	D	F	result
S_0:	OK	OK	OK	OK
S_1:	OK	OK	inactive	OK*
S_2:	OK	inactive	OK	OK*
S_3:	OK	inactive	inactive	OK*
S_4:	fails	OK	OK	fail-safe
S_5:	fails	OK	inactive	fail-dangerous
S_6:	fails	inactive	any	fail-dangerous

where * indicates that while the output is correct, some part of the system is not functioning correctly. The transitions between the states are shown in the Markov diagram of Figure 8.10.

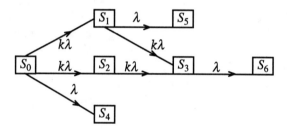

Figure 8.10. Markov diagram for Example 8.9

If $P_i(t)$ is the probability that the system is in state i at time t, then the reliability of the system is given by

$$R(t) = \sum_{i=0}^{3} P_i(t).$$

By the methods of Section 6.8 it can be shown that

$$R(t) = e^{-\lambda t}\left[e^{-k\lambda t} + \frac{1}{2} - \frac{1}{2}e^{-2k\lambda t}\right].$$

Of more importance here is the likelihood that, given a failure occurs, it is a dangerous failure. The probability of being in a fail-safe condition at time t is $\lambda e^{-\lambda(1+2k)t}$. So the probability that the failure is a fail-safe failure is

$$\frac{\lambda e^{-\lambda(1+2k)t}}{1-R(t)} = p_{fs}(t),$$

say, and the probability that the failure is fail-dangerous is $1 - p_{fs}(t)$.

Soper and Wolstenholme (1993) consider additional features of such a system. Unit D has several modes of failure. It may be active but fails to detect an error in M or it may falsely register an error. Also there is the facility for F to not only fail to overide M but also to itself fail dangerously.

The next example is based on a problem concerning the monitoring of potential failure in an atomic reactor, given by Thompson (1988).

Example 8.10: The rate of reaction in the atomic plant is controlled by rod insertion. *Transients*, the times at which control is needed, are assumed to occur according to a Poisson process, rate λ per reactor year. When a transient occurs the conditional probability that the safety system does not work and an accident results is q (very small, typically of the order 10^{-10}).

Suppose there are k transients in a reactor year. The probability that the safety system works every time is $(1 - q)^k$. The probability of no accident in a reactor year is given by

$$P = \sum_{k=0}^{\infty} [P(k \text{ transients}) \times P(\text{safety system works every time})]$$

$$= \sum_{k=0}^{\infty} \frac{\lambda^k e^{-\lambda}}{k!}(1-q)^k.$$

The probability that k transients occur is given by the Poisson distribution. Rewriting P in order to evaluate the sum of probabilities,

$$P = e^{-\lambda} \sum_{k=0}^{\infty} \frac{[\lambda(1-q)]^k}{k!} \quad \text{(see the Appendix for the series expansion of } e^x\text{),}$$

$$= e^{-\lambda}e^{\lambda(1-q)} = e^{-\lambda q}.$$

So the probability of at least one accident in a reactor year is $1 - e^{-\lambda q} = 1 - [1 - \lambda q + (\lambda q)^2/2! - \ldots]$, which is approximately λq for small λq.

Now suppose that inspection and repair of the safety system occurs at r regular scheduled times per year and let it be assumed that safety system failure goes undetected and unrepaired until inspection occurs. The time to safety system failure will be taken to be exponentially distributed, with parameter θ, and to be independent of transients.

Let the interval between safety system inspections be t_0. The probability that there is no safety system failure over a period of t_0 years is $e^{-\theta t_0}$. If the safety system failed at time t ($< t_0$), then no accident occurs in time t_0 if a transient does not occur in the interval $[t_0 - t]$. The probability of this combination of events is given by integrating the joint probability over t. (Note that it is not 'summing' this time because the variable t is continuous.)

The probability of no accident within t_0 years is given by

$$P(t_0) = e^{-\theta t_0} + \int_0^{t_0} [\theta e^{-\theta t}][e^{-\lambda(t_0 - t)}]\mathrm{d}t.$$

$$= \frac{\lambda e^{-\theta t_0} - \theta e^{-\lambda t_0}}{(\lambda - \theta)}.$$

If we take inspections to be regularly spaced, $t_0 = 1/r$ reactor years. The probability of no accident during the whole year is $[P(1/r)]^r$. So the annual probability of at least one accident is

$$Q_r = 1 - \left[P\left(\frac{1}{r}\right)\right]^r = 1 - \left[\frac{\lambda e^{-\theta/r} - \theta e^{-\lambda/r}}{\lambda - \theta}\right]^r$$

$$= 1 - \left(\frac{\lambda}{\lambda - \theta}\right)^r \left(e^{-\theta/r} - \frac{\theta e^{-\lambda/r}}{\lambda}\right)^r$$

$$= 1 - \left(1 - \frac{\theta}{\lambda}\right)^{-r} (e^{-\theta/r})^r \left(1 - \frac{\theta e^{(\theta - \lambda)/r}}{\lambda}\right)^r.$$

Applying Taylor expansions, we have

$$Q_r = 1 - \left(1 + \frac{r\theta}{\lambda} + \ldots\right)(1 - \theta + \ldots)\left(1 - \frac{r\theta e^{(\theta - \lambda)/r}}{\lambda} + \ldots\right).$$

Now θ will be very small, so ignoring all terms in θ of second order or higher,

$$Q_r \approx 1 - \left\{ 1 - \theta\left[-\frac{r}{\lambda} + 1 + \frac{r}{\lambda}e^{-\lambda/r} \right] \right\}$$

$$= \theta\left[1 + \frac{r}{\lambda}(e^{-\lambda/r} - 1) \right].$$

In the case where $\lambda = r$ (that is, inspections occur at intervals equal to the mean time to the occurrence of a transient) for example, then $Q_r \approx \theta[1 + (e^{-1} - 1)] = 0.368\theta$. Figure 8.11 demonstrates Q_r/θ as a function of r/λ.

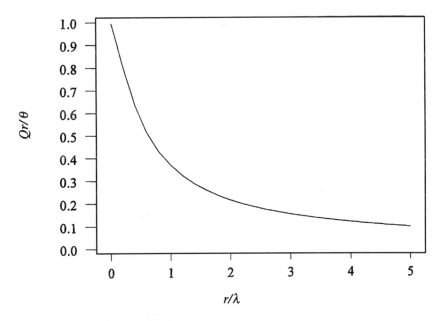

Figure 8.11. Relationship between annual probability of failure and frequency of inspection, scaled by the appropriate failure rate parameters

8.8   Down-time distributions

While there is often much evidence and reasoning for the assumption of an exponential model for lifetimes, the distribution of down time is frequently very clearly not exponential. One reason is that down time is generally composed of a number of stages: diagnosing the fault, obtaining parts and labour, physical repair time and possibly administrative/logistic phases. Each of these periods may have quite diverse time distributions. The total down time is unlikely in these circumstances to be exponentially distributed. Departure from the constant hazard rate assumption makes Markov, or other standard detailed analysis, complex, even for small systems. The lognormal distribution is often found to be a suitable down-time distribution, but it has the additional disadvantage of having a non-analytic distribution function. Long-term measures may still be simply calculated but short-term system behaviour would need often to be demonstrated via simulation.

Smith (1993) gives an extensive practical guide to the components of down time and the factors influencing the periods spent in each stage of down time. The maintenance strategy is highly influential. Allied to this are features of system design which should ideally make fault diagnosis and parts replacement as easy as possible. The technical competence required of staff and the support given, by maintenance manuals for example, are important considerations. Efforts to reduce down time once a system is in operation require detailed reporting of the different periods in the repair process, the time spent, the resources used and special factors.

Maintainability is generally taken to be the probability that a failed unit/system is restored to satisfactory working order within a certain period of time. It is not so much to do with how often the system goes down, which is a function of reliability, but how easy/difficult it is to repair the system. A highly reliable system may in some circumstances lose some of its appeal if it has low maintainability.

CHAPTER 9

Life Testing and Inference

9.1 Life test plans

The only way to measure reliability is to test completed products or components, under conditions that simulate real life, until failure occurs. Extensive testing, however, often results in undesirable expenditures of time and money.

Suppose that n units are placed on test and at the time the test is terminated there are r failures. This gives r uncensored observations and $n - r$ right-censored observations. If the test is stopped after a fixed length of time, so that the number of failures is random, then the data have *Type I censoring* (that is, time censoring). Let $t_{(1)}$, ..., $t_{(r)}$ be the first r observed lifetimes of n units on test. The *total time on test* is given by

$$\text{TT} = \sum_{i=1}^{r} t_{(i)} + (n - r)T_r, \qquad (9.1)$$

where T_r is the time to when the test is terminated.

If the test is stopped at the time of the rth failure, the number of failures is fixed but the test length is random, and the data have *Type II censoring* (or failure censoring). Testing may be done with or without *replacement*, that is, the experiment may have a certain number of test 'beds' and new units may or may not be replaced by new ones until r failures have occurred in total or the required testing time elapsed. For testing with replacement and Type II censoring, the duration of the test, T_r, is shorter than without replacement. For testing without replacement and Type II censoring, $T_r = t_{(r)}$, that is, $E[T_r] = E[t_{(r)}]$. If all units have exponential lifetimes with rate parameter λ,

$$E[t_{(r)}] = \frac{1}{\lambda}\left[\frac{1}{n} + \frac{1}{n-1} + \frac{1}{n-2} + \dots + \frac{1}{n-(r-1)}\right] = \frac{1}{\lambda}\sum_{i=1}^{r}\frac{1}{n+1-i}.$$

For testing with replacement and Type II censoring,

$$E[T_r] = \frac{1}{\lambda}\sum_{i=1}^{r}\frac{1}{n} = \frac{r}{\lambda n},$$

but $n + r - 1$ units are placed on test and the number and proportion of censored observations is higher.

Sequential plans are 'accept–reject' tests under a given null hypothesis, H_0, versus an alternative hypothesis H_1. The life test is continuously monitored and a decision made as soon as there is sufficient supporting evidence for one of the two hypotheses. These tests take less time than non-sequential plans but estimation is complicated and not very 'robust'.

Various factors influence the choice of test plans, usually in relation to resources. These may be physical, time-related or financial.

Example 9.1: Let there be n units on test originally and the experiment terminated after r failures. Suppose there to be a set-up cost of nC_1, a running cost of C_2 per unit time, and a replacement cost of C_3 per item. Of interest is the total cost, C, of conducting the experiment. If failed units are replaced, the experiment is likely to take less time but will involve replacement costs. The expected total cost is

$$E(C) = nC_1 + C_2\, E(T_r) + mC_3,$$

where m is the number of items replaced. Without replacement, $m = 0$ and with replacement, $m = r - 1$. The optimum cost of the experiment is to do with how $E(T_r)$ varies under the two testing strategies and also the relative sizes of C_2 and C_3. It may be, however, that cost is not the only consideration. The value of T_r may be important and this may lead to also varying n in order to achieve the most generally effective experiment.

Example 9.2: Consider a non-replacement plan with Type II censoring for screening consignments of a particular electrical device. Suppose n^*, the value of n which minimizes the cost of the experiment for given r, is required. It will be assumed that the devices have exponential lifetime, with rate parameter λ per hour.

Now $E(C) = nC_1 + C_2\, E(T_r)$ and

$$E(T_r) = \frac{1}{\lambda}\sum_{i=1}^{r}\frac{1}{n+1-i}.$$

Clearly n^* is dependent on λ and the relative sizes of C_1 and C_2.

As an illustration, λ will be taken to be 0.01 and C_1/C_2 will be taken to be 20. Minimizing $E(C)$ is then equivalent to minimizing $f(n) = 20n + E(T_r)$. Figure 9.1 shows that a practical value for n^* is 13. Further similar calculations can show how sensitive n^* is to the value chosen for λ, and of course to the lifetime model assumed.

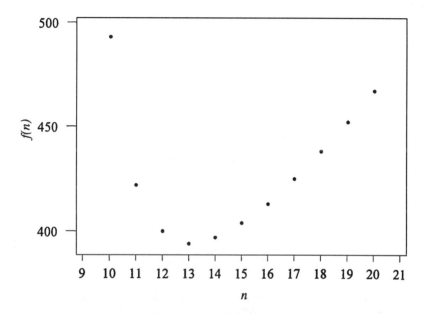

Figure 9.1. Cost function $f(n)$ for Example 9.2

9.2   Prediction of time on test

Having collected some test data, estimates of future life or the time needed to collect a certain number of further observations may well be of interest.

Suppose a sample of n observations $X_1, X_2, \ldots, X_n$, has mean $\overline{X}$ and is independent of a future sample of m observations, $T_1, T_2, \ldots, T_m$. We could be interested in $T_{(1)}$, the smallest observation of a future sample or the life of a series system of m components; $T_{(m)}$, the life of a parallel system of m components or the time to complete a test of m units placed on test simultaneously; or $\text{TT}_m = \sum_{i=1}^{m} T_{(i)}$, the total test time for m more failures.

Unconditionally, an unbiased predictor for the future mean $\bar{T}$ is $\bar{X}$. TT_m is estimated by $m\bar{X}$. Note that, in general, if a statistic $\hat{d}$ is used to predict d, the *prediction error* is $\hat{d} - d$.

Exponential data from a population with rate parameter λ

For the prediction error, $\bar{X} - \bar{T}$, the variance is

$$V(\bar{X} - \bar{T}) = V(\bar{X}) + V(\bar{T})$$

$$= V\left(\frac{X_1 + X_2 + \ldots + X_n}{n}\right) + V\left(\frac{T_1 + T_2 + \ldots + T_m}{m}\right)$$

$$= \frac{V(X)}{n} + \frac{V(T)}{m} \quad \text{(from (7.3) and (7.4))}$$

$$= \frac{1/\lambda^2}{n} + \frac{1/\lambda^2}{m} = \frac{m+n}{\lambda^2 mn}.$$

It can be shown that two-sided prediction limits for TT_m are approximately given by

$$\left[\frac{m\bar{X}}{F_{(2n, 2m)}}, m\bar{X}F_{(2m, 2n)}\right],$$

where $F_{(v_1, v_2)}$ is the appropriate percentage point from the F distribution with (v_1, v_2) degrees of freedom. These points are tabulated in standard statistical tables. (Note that the lower $(\alpha/2)\%$ point on $F_{(v_1, v_2)}$ is the reciprocal of the $100 - (\alpha/2)\%$ point on $F_{(v_2, v_1)}$.)

An unbiased predictor for $T_{(1)}$ is $\bar{X}/m$, and two-sided prediction limits are given by

$$\frac{\bar{X}}{m}\left[\frac{1}{F_{(2n, 2)}}, F_{(2, 2n)}\right].$$

A lower limit for $T_{(1)}$ could be interpreted as a 'safe warranty life'.

One-sided prediction intervals may be constructed in the usual way.

Normal data (or log data from a lognormal distribution)

Two-sided prediction limits for the mean $\overline{T}$ of a future sample of size m based on a previous sample of size n with mean $\overline{X}$ and standard deviation S are given by

$$\overline{X} \pm t_{(n-1)} S \sqrt{\frac{1}{m} + \frac{1}{n}}$$

where $t_{(n-1)}$ is the appropriate percentage point from the t distribution with $n-1$ degrees of freedom.

The following (biased) predictors are used for $T_{(1)}$ and $T_{(m)}$:

$$\hat{T}_{(1)} = \overline{X} - z_c\left(\frac{m - 0.5}{m}\right)S,$$

$$\hat{T}_{(m)} = \overline{X} + z_c\left(\frac{m - 0.5}{m}\right)S.$$

Example 9.3: The following data from Nelson (1982) are lifetimes (in hours) for electrical insulation at 190°C:

7228, 7228, 7228, 8449, 9166, 9166, 9166, 9166, 10510, 10510.

The coarseness of the data is due to failure being registered periodically. The log data are close to normal, implying that the lifetimes are lognormal. The mean and standard deviation of the log data are respectively 9.0712 and 0.1439. A further sample of $m = 10$ specimens is to be tested at 190°C. The prediction of the log life of the longest-running specimen is

$$\hat{T}_{(10)} = 9.0712 + z\left(\frac{10 - 0.5}{10}\right)0.1439 = 9.3079.$$

Therefore exp(9.3079) = 11025 hours is a prediction of the time required to run all specimens to failure when they are all put on test at the same time.

The effect of coarsely recording data will be discussed in Section 9.4.

Weibull and extreme value (Gumbel) data

The most efficient methods in this case are based on best linear unbiased estimators (BLUEs) and require detailed computation. Mann *et al.* (1974) provide a comprehensive account.

The smallest future observation, $T_{(1)}$, from a sample of size m may be simply predicted using the approximate estimation of the Gumbel distribution parameters in (3.11):

$$\hat{T}_{(1)} = \hat{\gamma} - \hat{\eta}\log(m).$$

This predictor estimates the mode of the distribution of $\hat{T}_{(1)}$ and is biased downwards because the distribution of $\hat{T}_{(1)}$ has a long right-hand tail.

9.3   Inference for the exponential distribution

For lifetimes with probability density function $f(x)$ and distribution function $F(x)$, the likelihood function, where there are r observed lifetimes and $n - r$ right-censored lifetimes, is given by

$$L(\lambda) = \left\{\prod_{i=1}^{r} f(t_{(i)})\right\}[1 - F(t_{(r)})]^{n-r}. \qquad (9.2)$$

For exponential lifetimes, $L(\lambda) = \lambda^r \exp(-\lambda T)$, where T is the sum of all the uncensored and censored lifetimes, that is, the total time on test.

Following Example 4.2, we find that the maximum likelihood estimate of λ is

$$\hat{\lambda} = r/T. \qquad (9.3)$$

Letting

$$W_i = (n + 1 - i)(t_{(i)} - t_{(i-1)}),$$

and noting that $t_{(0)} = 0$, then

$$T = \sum_{i=1}^{r} W_i .$$

Each W_i is the total time on test between the $(i-1)$th and the ith failures. It can be shown that the $\{W_i\}$ are independent random variables, each with probability density function $\lambda \exp(-\lambda w)$, that is, have the same distribution as the $T_{(i)}$. This arises because, due to the Markov (memoryless) property of the exponential distribution, it does not matter whether the total test time to the next failure is spread over one or several similar Poisson processes. It then follows that $2\lambda T$ has a chi-squared distribution with $2r$ degrees of freedom $(\chi^2_{(2r)})$, which provides the basis of confidence intervals for λ and also for functions of λ, such as the reliability.

The confidence interval for λ may be described in general terms as $[\lambda_L, \lambda_U]$. At $100(1-\alpha)\%$ confidence,

$$\lambda_L = \frac{\chi^2_{(2r;\alpha/2)}}{2T}, \ \lambda_U = \frac{\chi^2_{(2r;1-\alpha/2)}}{2T}. \tag{9.4}$$

A confidence interval for the mean is given by

$$1/\lambda_U \le 1/\lambda \le 1/\lambda_L.$$

A confidence interval for the reliability is given by

$$e^{-\lambda_U t} \le R(t) \le e^{-\lambda_L t}. \tag{9.5}$$

A confidence interval for the pth quantile is given by

$$\frac{-\log(1-p)}{\lambda_U} \le t_p \le \frac{-\log(1-p)}{\lambda_L}.$$

Example 9.4: The first eight observations (in hours) in a random sample of size 12 from an assumed exponential population are

$$31, 58, 157, 185, 300, 470, 497, 672.$$

So $n = 12$, $r = 8$, $T = 5058$ (taking into account the four observations right-censored at 672 hours). For a χ^2 distribution with 16 degrees of freedom, the 2.5% and 97.5% points are 6.91 and 28.8 respectively, so from (9.4) we have a 95% confidence interval for λ given by

$$\frac{6.91}{2 \times 5058} \leq \lambda \leq \frac{28.8}{2 \times 5058}$$

$$0.000683 \leq \lambda \leq 0.00285.$$

The maximum likelihood estimate of λ is $8/5058 = 0.00158$, which, it is noted, is not the mid-point of the above interval because the distribution of the estimator is not symmetrical. This should be contrasted with the normal approximation used in Section 4.4.

The 95% point of $\chi^2_{(16)}$ is 26.3, so a lower 95% confidence limit for the mean lifetime is $1/\lambda_U$, where λ_U is $26.3/(2 \times 5058) = 0.0026$. So $1/\lambda_U = 384.6$ hours.

The lifetime exceeded by 99% of lives is given by $t_{0.01}$, and a lower 95% confidence limit for this quantile is given by

$$\frac{-\log(1-0.01)}{\lambda_U} = \frac{0.01005}{0.0026} = 3.865 \text{ hours.}$$

It is informative to note how low is this bound for $t_{0.01}$. It is a function of the constant hazard of the exponential model, and of the sample size, and illustrates how deficient the model might be in practical use.

Type I censored data are often analysed as though Type II. This is often reasonably satisfactory, though it can be readily appreciated that the total time on test for r failures is now higher than under Type II censoring. A simple approximate correction may be applied whereby $\chi^2_{(2r+1)}$ is used as the sampling distribution rather than $\chi^2_{(2r)}$.

Example 9.5: Suppose the test of Example 9.4 was terminated at 720 hours rather than 672, with still the same number of recorded failures. The total time on test is now 5250 and interval estimates will be based on $\chi^2_{(17)}$. The maximum likelihood estimate of λ is $8/5250 = 0.001524$, lower than before because we have had a longer testing time with the same number of failures, and this little extra information also slightly shortens the 95% interval estimate,

$$\frac{7.564}{2 \times 5250} \leq \lambda \leq \frac{30.19}{2 \times 5250}$$

$$0.000720 \leq \lambda \leq 0.00287.$$

9.4   The effect of data rounding

Lifetime data often arise as a result of inspection at intervals (see Example 9.3). It may be known that a unit is working at time $t - h$, but that it is not working at time t. In the context of conducting a life test, the effect of the choice of inspection interval may be readily calculated. As an example, consider lifetimes thought to be exponential and when recorded at t, actually lie in the interval $(t - h, t)$. Times which are right-censored can be taken to be exactly t. We have now effectively interval-censored observations for actual lifetimes and the likelihood function of (9.2) becomes

$$L(\lambda) = \left[\prod_{i=1}^{r}\{F(t_{(i)}) - F(t_{(i)} - h)\}\right][1 - F(t_{(r)})]^{n-r}$$

$$= \left[\prod_{i=1}^{r}\{e^{-\lambda t} - e^{-\lambda(t_{(i)} - h)}\}\right][e^{-\lambda t_{(r)}}]^{n-r}$$

$$= (e^{\lambda h} - 1)^{r}\left[\prod_{i=1}^{r}e^{-\lambda t_{(i)}}\right]e^{-\lambda(n-r)t_{(r)}}.$$

So the log likelihood is

$$\log L(\lambda) = r\log(e^{\lambda h} - 1) - \lambda T.$$

$$\frac{\partial \log L}{\partial \lambda} = \frac{rhe^{\lambda h}}{e^{\lambda h} - 1} - T$$

The maximum likelihood estimate of λ is given by setting this derivative to zero, and this yields

$$\hat{\lambda} = \frac{1}{h}\log\left(\frac{T}{T - rh}\right).$$

The estimated variance of $\hat{\lambda}$ is given (see Section 4.4) by

$$\left(-\frac{\partial^{2}\log L}{\partial \lambda^{2}}\right)^{-1} = \frac{(e^{\lambda h} - 1)^{2}}{rh^{2}e^{\lambda h}}, \text{ when } \lambda \text{ is replaced by } \hat{\lambda}.$$

Example 9.6: Suppose the data of Example 9.4 were recorded rounded up to the nearest day, as may be the case if inspection takes place once at the same time each day. The data would be as follows:

$$2, 3, 7, 8, 13, 20, 21, 28,$$

and the right-censored observations would all be at 28 days.

We now have $T = 214$, $h = 1$ and $r = 8$. If the rounding is ignored, $\hat{\lambda} = r/T = 0.03738$, with estimated $V(\hat{\lambda}) = \hat{\lambda}^2/r = 1.7469 \times 10^{-4}$.

Incorporating the rounding yields

$$\lambda = \frac{1}{1}\log\left(\frac{214}{214-8}\right) = 0.03810$$

with estimated $V(\hat{\lambda}) = 1.8145 \times 10^{-4}$.

In the working for Example 9.4, λ is estimated to be 8/5058 per hour. So converting to 'per day' by multiplying by 24 gives $\hat{\lambda} = 0.03796$ compared to 0.03738 above.

A simple approach to concerns about rounding is to site uncensored observations half-way between inspection points – in this case, therefore, to subtract 0.5 from each observation. The value of T is now 210 and $\hat{\lambda} = r/T = 0.038095$, very close to that given by the interval-censored approach. The effect of this strategy is model-dependent. The approximation will work to different degrees of accuracy for different lifetime models. However, the central message is that there is little to be gained from high degrees of recording accuracy. A common-sense approach will generally suffice.

9.5   Parametric reliability bounds

Example 9.7: Example 4.5 constructed exponential reliability bounds using an approximate normal distribution for the estimate of λ. Using that example and (9.5) would yield a lower 95% bound for the reliability

$$\exp\left\{-\left(0.08 + 1.645\frac{0.08}{\sqrt{8}}\right)t\right\} = \exp\{-0.1265t\}.$$

Using the exact theory of Section 9.3 the 95% upper estimate for λ is

$$\lambda_U = \frac{26.3}{2 \times 100} = 0.1315,$$

so giving a lower 95% reliability bound of $\exp\{-0.1315t\}$.

The resulting differences in specific bounds are illustrated below:

t	normal approx., $R = e^{-0.1265t}$	exact, $R = e^{-0.1315t}$
0.5	0.9387	0.9364
1	0.8812	0.8768
5	0.5313	0.5181

In practice it is the lower bounds which are of maximum interest and these are slightly overestimated (optimistic) if the normal approximation is used.

Where a reliability function has more than one estimated parameter, a reliability bound is unlikely to be a simple function of parameter bounds.

Example 9.8: The Gumbel reliability function is given by

$$R(y) = \exp\left[-\exp\left(\frac{y-\gamma}{\eta}\right)\right],$$

and bounds are given by $\exp[-\exp(c_U)]$, $\exp[-\exp(c_L)]$ where c_L, c_U are the lower and upper confidence limits of $\xi = (y - \hat{\gamma})/\hat{\eta}$. These will depend on both the parameter point estimates and their variances.

If we have uncensored data and use the methods of Section 4.6, we have approximately normally distributed $\hat{\gamma}$ and $\hat{\eta}$ with variances $1.168\hat{\eta}^2/n$ and $1.1\hat{\eta}^2/n$, respectively. From (7.6)

$$V(\xi) \approx \left\{\frac{\partial \xi}{\partial \gamma}\bigg|_{\gamma = \hat{\gamma}, \eta = \hat{\eta}}\right\}^2 V(\hat{\gamma}) + \left\{\frac{\partial \xi}{\partial \eta}\bigg|_{\gamma = \hat{\gamma}, \eta = \hat{\eta}}\right\}^2 V(\hat{\eta}).$$

Now

$$\frac{\partial \xi}{\partial \gamma} = -\frac{1}{n} \quad \text{and} \quad \frac{\partial \xi}{\partial \eta} = -\frac{1}{\eta}\left(\frac{y-\gamma}{\eta}\right).$$

So

$$V(\xi) \approx \frac{1}{\hat{\eta}^2}\left\{\frac{1.168\hat{\eta}^2}{n} + \frac{1}{\hat{\eta}^2}\xi^2\frac{1.1\hat{\eta}^2}{n}\right.$$

$$= \frac{1}{n}[1.168 + 1.1\xi^2].$$

The limits (c_U, c_L) are then given by

$$\xi \pm z_{c}\sqrt{\frac{1.168 + 1.1\xi^2}{n}}.$$

The formula for $V(\xi)$ may be adjusted to take account of the slight negative covariance between the parameter estimates and one such adjustment is illustrated by Nelson (1982, p. 234). Here the term 0.1913ξ is subtracted from the above $V(\xi)$. A different, more complex approach is shown in Section 9.6.

Example 9.9: The Weibull reliability function is given by $R(x) = \exp\{-(x/\alpha)^\beta\}$. We can proceed as above to calculate bounds for the exponent or we can use the reliability function itself.

From (7.10),

$$V(R) \approx \left\{\frac{\partial R}{\partial \alpha}\bigg|_{\alpha=\hat{\alpha}, \beta=\hat{\beta}}\right\}^2 V(\hat{\alpha}) + \left\{\frac{\partial R}{\partial \beta}\bigg|_{\alpha=\hat{\alpha}, \beta=\hat{\beta}}\right\}^2 V(\hat{\beta})$$

$$= \hat{R}^2\left\{[x^{\hat{\beta}}\hat{\beta}\hat{\alpha}^{-\hat{\beta}-1}]^2 V(\hat{\alpha}) + \left[-\left(\frac{x}{\hat{\alpha}}\right)^{\hat{\beta}}\log\left(\frac{x}{\hat{\alpha}}\right)\right]^2 V(\hat{\beta})\right\}$$

$$= \hat{R}^2\left(\frac{x}{\hat{\alpha}}\right)^{2\hat{\beta}}\left\{\frac{\hat{\alpha}^2}{\hat{\beta}^2}V(\hat{\alpha}) + \left[\log\left(\frac{x}{\hat{\alpha}}\right)\right]^2 V(\hat{\beta})\right\}.$$

Again it should be noted that an upper bound for R cannot be obtained by simply substituting, say, the upper bound for α and the lower bound for β.

We can use a variety of estimates and their variances in this expression but the most satisfactory approach is via maximum likelihood.

9.6   Likelihood-based methods

Maximum likelihood methods are very important because they possess good properties and are extremely versatile. In particular, they are applicable to most types of censored data.

The asymptotic properties of the likelihood function allow confidence intervals or regions to be calculated under an assumed parametric model. This usually involves one of the following:

(i)   the approximate normal distribution of maximum likelihood estimates;

(ii) the approximate χ^2 distribution of the *likelihood ratio* statistic.

Methods involving (ii) are likely to be better than those involving (i) with smaller samples, but (i) is easier to use. Sometimes a transformation of the maximum likelihood estimator, such as the log or square root, has a more closely normal distribution. Occasionally it is practical to use the exact distribution of the estimator, as in the exponential case of Section 9.3. A confidence interval for parameter θ_i will take the form of (4.5), where the variance of the estimator is found from the second derivatives of the log-likelihood function.

Example 9.10: Consider the Weibull distribution with shape parameter β and scale parameter α. If the random variable T has a Weibull distribution, then $X = \log T$ has a Gumbel distribution with location parameter $\gamma = \log \alpha$ and scale parameter $\eta = 1/\beta$. It is often easier to work with the Gumbel distribution and then translate the results back into the Weibull distribution.

Let there be n observations, r of which form a set U of uncensored lifetimes. The log likelihood function is

$$\log L(\gamma, \eta) = -r\log \eta + \sum_{i \in U} \frac{x_i - \gamma}{\eta} - \sum_{i=1}^{n} \exp\left(\frac{x_i - \gamma}{\eta}\right),$$

where the x_i are logs of the lifetimes.

Differentiating with respect to γ gives

$$\frac{\partial \log L}{\partial \gamma} = \sum_{U} \left(\frac{-1}{\eta}\right) - \sum_{i=1}^{n} \exp\left(\frac{x_i - \gamma}{\eta}\right)\left(\frac{-1}{\eta}\right)$$

which is zero when

$$r = \sum_{i=1}^{n} \exp\left(\frac{x_i - \gamma}{\eta}\right) = \frac{\sum_{i=1}^{n} \exp(x_i/\eta)}{\exp(\gamma/\eta)} .$$

This gives

$$\hat{\gamma} = \hat{\eta}\log\left\{\frac{1}{r}\sum_{i=1}^{n} \exp(x_i/\hat{\eta})\right\} .$$

Differentiating $\log L$ with respect to η gives

$$\frac{\partial \log L}{\partial \eta} = -\frac{r}{\eta} + \sum_{U} \frac{-(x_i - \gamma)}{\eta^2} - \sum_{i=1}^{n} \exp\left(\frac{x_i - \gamma}{\eta}\right)\left[\frac{-(x_i - \gamma)}{\eta^2}\right] .$$

Setting $\partial \log L/\partial \eta = 0$ and substituting for $\exp(\hat{\gamma}/\hat{\eta})$ from the equation for $\hat{\gamma}$, we have

$$0 = -r\hat{\eta} - \sum_{U} x_i + r\hat{\gamma} + r\frac{\sum_i x_i \exp(x_i/\hat{\eta})}{\sum_i \exp(x_i/\hat{\eta})} - r\hat{\gamma} .$$

The maximum likelihood estimate, $\hat{\eta}$, of η is found by solving iteratively the equation

$$\frac{\sum_i x_i \exp(x_i/\hat{\eta})}{\sum_i \exp(x_i/\hat{\eta})} - \hat{\eta} - \frac{1}{r}\sum_{U} x_i = 0 .$$

A good initial estimate may be obtained graphically or using (4.7), treating all data as uncensored. The value of $\hat{\gamma}$ is then yielded directly by substitution.

A number of statistical packages, MINITAB for example, will yield maximum likelihood estimates of Weibull parameters, but may only do so for uncensored data and may not give standard errors of the estimates.

Now letting $z_i = (x_i - \hat{\gamma})/\hat{\eta}$, the observed information matrix I_0 takes a very simple form.

$$I_0 = \begin{bmatrix} -\dfrac{\partial^2 \log L}{\partial \gamma^2} & -\dfrac{\partial^2 \log L}{\partial \gamma \partial \eta} \\[2mm] -\dfrac{\partial^2 \log L}{\partial \gamma \partial \eta} & -\dfrac{\partial^2 \log L}{\partial \eta^2} \end{bmatrix}$$

$$= \frac{1}{\hat{\eta}^2} \begin{bmatrix} r & \displaystyle\sum_{i=1}^{n} z_i \exp z_i \\[3mm] \displaystyle\sum_{j=1}^{n} z_i \exp z_i & r + \displaystyle\sum_{i=1}^{n} z_i^2 \exp z_i \end{bmatrix}.$$

Taking $(\hat{\gamma}, \hat{\eta})$ to be approximately bivariate normally distributed with mean (γ, η) and covariance matrix I_0^{-1}, given by

$$I_0^{-1} = \begin{bmatrix} V(\hat{\gamma}) & C(\hat{\gamma}, \hat{\eta}) \\ C(\hat{\gamma}, \hat{\eta}) & V(\hat{\eta}) \end{bmatrix}$$

confidence intervals for γ and η, and quantiles for example, may be constructed.

The pth quantile of the distribution of X is given by $x_p = \gamma + w_p \eta$, where $w_p = \log[-\log(1-p)]$. Let x_p be estimated by $\hat{x}_p = \hat{\gamma} + w_p \hat{\eta}$. Assuming $\hat{x}_p$ to be approximately normally distributed, a confidence interval for x_p is given by

$$\hat{x}_p \pm z_c \sqrt{V(\hat{x}_p)} \ ,$$

where

$$V(\hat{x}_p) = V(\hat{\gamma} + w_p \hat{\eta})$$
$$= V(\hat{\gamma}) + 2C(w_p \hat{\eta}, \hat{\gamma}) + V(w_p \hat{\eta}) \quad \text{(see (7.6))}$$
$$= V(\hat{\gamma}) + 2w_p C(\hat{\eta}, \hat{\gamma}) + w_p^2 V(\hat{\eta}).$$

Example 9.11: The data of Example 3.4 in Mann and Fertig (1973) give the results of a life test terminated after the tenth failure in a sample of 13. It has already been shown in Figure 3.4 that these data conform well to a Weibull distribution. The maximum likelihood estimates of the Weibull parameters α and β are respectively 2.27 and 1.42, hence the Gumbel distribution for the log lifetimes has parameters with maximum likelihood estimates $\hat{\gamma} = \log \hat{\alpha} = 0.820$ and $\hat{\eta} = 1/\hat{\beta} = 0.704$. Transforming the log data to z_i values then yields

$$I_0 = \frac{1}{(0.704)^2} \begin{bmatrix} 10 & 0.14218 \\ 0.14218 & 10 + 3.5957 \end{bmatrix},$$

so that

$$I_0^{-1} = \begin{bmatrix} 0.049565 & -0.00052 \\ -0.00052 & 0.036458 \end{bmatrix}.$$

Consider the quantile $x_{0.1}$, the log lifetime exceeded 90% of the time. This is estimated by

$$\hat{x}_{0.1} = \hat{\gamma} + \log(-\log(0.9))\hat{\eta}$$
$$= 0.820 + -2.25 \times 0.704 = -0.764,$$

giving the corresponding estimated lifetime quantile $\hat{t}_{0.1} = \exp(-0.764) = 0.466$. A 95% lower bound for $x_{0.1}$ is given by

$$-0.764 - 1.645\sqrt{0.049565 + 2(-2.25)(-0.00052) + (-2.25)^2 0.036458} = -1.564$$

The corresponding lifetime is $\exp(-1.564) = 0.209$.

9.7   The likelihood ratio test

Let $\boldsymbol{\theta}$ be a vector of parameters with set of possible values Θ. A pair of composite hypotheses may be expressed as

$$H_0: \boldsymbol{\theta} \in \Theta_0, \; H_1: \boldsymbol{\theta} \in \Theta_1,$$

where Θ_0 and Θ_1 are two disjoint subsets of Θ. Frequently $\Theta_1 = \Theta - \Theta_0$.

Under H_0, $\theta_i = \theta_{i0}$, where θ_{i0} may or may not be a numerical value. The hypothesis may specify particular values for some parameters and leave others to vary freely, or some relationship between parameters may be specified, for example $\theta_1 = \theta_2$.

A ratio of likelihoods is constructed,

$$\lambda = \frac{L(\hat{\theta}_0)}{L(\hat{\theta})}, \tag{9.6}$$

where the denominator is the maximized likelihood function with respect to all parameters θ, and the numerator is the maximum only after some or all of the parameters have been restricted by H_0.

$L(\hat{\theta}_0)$ cannot exceed $L(\hat{\theta})$, so λ lies between 0 and 1. The closer it is to 1 the greater the belief in H_0 and the closer to zero the more inclined we are to reject H_0. Since λ is purely a function of the observed data $\{x_{ij}\}$ it is a random variable, but in order to be able to use λ as a statistic for testing H_0 the distribution of λ needs to be known. If λ is a simple function, the sample mean, $\bar{x}$, for example, its distribution may be found explicitly, but in general λ is not a simple function and its distribution hard to find exactly. However, for large samples, an approximate distribution is available.

The likelihood ratio statistic is

$$\Lambda = -2\log\lambda = 2[\log(\hat{\theta}) - \log(\hat{\theta}_0)]. \tag{9.7}$$

For large samples, Λ has an approximate χ^2 distribution with degrees of freedom given by the number of parameters which are estimated under H_1 but fixed under H_0. Small values of λ correspond to large values of Λ, so values in the right-hand tail of χ^2 the distribution count against H_0. The test can be shown to possess some very desirable properties and many statistical tests in common use are in fact likelihood ratio tests.

The likelihood ratio statistic can be used to construct confidence regions for parameter estimates. In the case of a single parameter distribution, such as the exponential, the likelihood L_1 is the value of the likelihood function when the parameter θ is replaced by its maximum likelihood estimate. A confidence interval for θ may be constructed from the range of alternative values for θ which produce a likelihood L_0 such that the difference between $2\log L_0$ and $2\log L_1$, the likelihood ratio statistic Λ, is less than the appropriate percentage point of $\chi^2_{(1)}$. For a 95% confidence interval for θ the value from $\chi^2_{(1)}$ is 3.84. Figure 9.2 illustrates.

Extending this principle to distributions with more than one parameter, a *confidence region* for the Weibull parameters (α, β) is given by all (α_0, β_0) such that Λ is less than the appropriate percentage point of $\chi^2_{(2)}$.

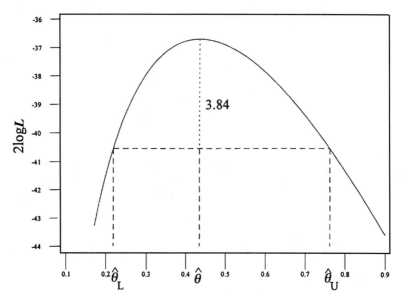

Figure 9.2. Interval estimation from the likelihood function

Example 9.12: Wolstenholme (1991) examines the strength characteristics of tungsten and silicon carbide fibres. A Weibull model is considered
for data collected from five different material structures, N, O, P, Q, R.
The data sets may be compared by considering the degree to which
confidence regions for the model parameters do or do not overlap. No
overlap implies a significant difference in model parameters, but some
overlap does not necessarily imply no significant difference. A 95% con-
fidence region is defined by (α_0, β_0) such that $2\log L(\alpha_0, \beta_0)$ is within 5.99
of $2\log L(\hat{\alpha}, \hat{\beta})$, where 5.99 is the upper 5% point of $\chi^2_{(2)}$. Figure 9.3 shows
that the strengths of N, O, and P are possibly similar, but quite distinct
from Q and R, which are similar to each other.

Confidence regions for Weibull parameters are characteristically
rather 'banana-shaped'. It can be more satisfactory to use logged data
and the Gumbel distribution. Confidence regions for Gumbel parameters are more elliptical in shape.

The use of the likelihood ratio test is confined to discrimination
within a given general model family. The hypotheses are about plausible parameter values, not about model suitability as such. The question of model comparison was discussed in Section 4.4. While a model
which yields the largest likelihood for a given set of data is the basic
principle, formal tests between model families are limited.

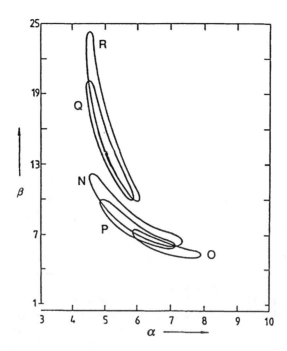

Figure 9.3. 95% confidence regions for Weibull parameters of Example 9.12

Example 9.13: The data of Example 9.11 may lead to the question of whether an exponential model would be an adequate alternative to the Weibull. In this case L_1 is the likelihood maximized with respect to α and β, and L_0 is the likelihood maximized with respect to α only and with β set to 1. Evidence that a Weibull model is more appropriate than the exponential would be given by Λ greater than $\chi^2_{(1)}$. Now

$$
\begin{aligned}
\log L_1 &= r\log\hat{\beta} + (\hat{\beta} - 1)\sum_{i=1}^{n}\log(t_i) - \sum_{i=1}^{n}\left(\frac{t_i}{\alpha}\right)^{\hat{\beta}} - r\hat{\beta}\log\hat{\alpha} \\
&= 10\log(1.42) + 0.42 \times 4.536 - 10.02 - 10 \times 1.42 \times 2.27 \\
&= -16.25.
\end{aligned}
$$

When β is set to 1, the log likelihood is equivalent to that for the exponential model with $\lambda = 1/\alpha$. The maximum likelihood estimate of λ is $\hat{\lambda} = r/(\sum t_i) = 10/23.05$. So,

$$\log L_o = -\frac{1}{\hat{\alpha}} \sum_{i=1}^{n} t_i - r\log\hat{\alpha}$$

$$= -\frac{10}{23.05} 23.05 - 10\log\left(\frac{23.05}{10}\right) = -18.35.$$

The likelihood ratio statistic $\Lambda = 2\log L_1 - 2\log L_0 = -32.5 + 36.7 = 4.2$. The degrees of freedom for the distribution of Λ are $2 - 1 = 1$. The 95% point of $\chi^2_{(1)}$ is 3.84 and since 4.2 is greater than 3.84 there is, at a 5% level of significance, evidence that for these data an exponential model is not an adequate alternative to the Weibull model.

Example 9.14: Use of the likelihood ratio test to examine the weakest-link hypthesis, demonstrated in Example 2.2, is found in Watson and Smith (1985). Weibull distributions are fitted to strength data for fibres at each of four different lengths. This involves a total of eight different Weibull parameters. If the weakest-link relationship and Weibull model hold for these fibres over the different lengths, then a single Weibull distribution would describe all strengths, with the size (length) characteristic represented by a defined parameter, as in Section 2.2, where a system of a series of n components is described as having 'size' n. Under H_1, the maximized log likelihood is calculated for data at each length and

$$\log L_1 = \log L_{11}(\alpha_1, \beta_1) + \log L_{12}(\alpha_2, \beta_2) + \log L_{13}(\alpha_3, \beta_3) + \log L_{14}(\alpha_4, \beta_4).$$

The weakest-link relationship defines H_0: $\alpha_1 = \alpha_2 = \alpha_3 = \alpha_4 = \alpha$ and $\beta_1 = \beta_2 = \beta_3 = \beta_4 = \beta$. The maximized log likelihood under H_0 is then calculated combining all data under a common model

$$f(x) = \frac{\beta}{\alpha_n}\left(\frac{x}{\alpha_n}\right)^{\beta-1} \exp\left[-\left(\frac{x}{\alpha_n}\right)^{\beta}\right],$$

where α_n is $\alpha_n^{-1/\beta}$. A summary of the results is shown below:

length (n)	α	β	$\log L$
1	4.58	5.6	−71.0
2	5.25	5.0	−63.0
3	4.57	5.5	−49.9
4	4.65	6.0	−36.2
total			−220.1
all	4.77	5.58	−229.1

The likelihood ratio statistic $\Lambda = -2\log\lambda = 2(-220.1 + 229.1) = 18.0$. The χ^2 distribution against which this value is compared has $8 - 2 = 6$ degrees of freedom. This is the difference between the number of parameters estimated under the different hypotheses. The upper 1% point of $\chi^2_{(6)}$ is 16.81. As a result, Watson and Smith conclude that for these fibres the weakest-link hypothesis, as represented by the Weibull distribution, does not hold.

There were between 57 and 70 observations at each length, so this is a fairly extensive dataset. The individual estimates for α and β appear similar, but the standard errors are small and the confidence regions do not coincide sufficiently to support a common model.

9.8   Binomial experiments

Let θ be the probability that a unit has a certain characteristic – for example, that it fails. Suppose units to be independent and θ constant from unit to unit. Recalling (6.6), for a sample of size n taken from an infinite population, the number of failed units, X say, will have a *binomial distribution*, $B(n,\theta)$, with probabilities given by

$$P(X = x) = \frac{n!}{(n-x)!x!}\theta^x(1-\theta)^{n-x}. \tag{9.8}$$

The mean of X is $n\theta$ and the variance $n\theta(1-\theta)$. The sample proportion of failures is $X/n = P$ and has mean θ and variance $\theta(1-\theta)/n$.

Confidence intervals for unknown θ

For p, an observed value of P, a confidence interval for θ may be constructed from the set of distributions $B(n, \theta)$ for which

$$P(X \geq x) = \alpha \quad \text{and} \quad P(X \leq x) = \alpha$$

where $x = np$ and α depends on the confidence level – for example, $\alpha = 0.025$ for a 95% confidence interval. A lower limit for θ is the solution to the equation

$$\sum_{r=x}^{n}\binom{n}{r}\theta^r(1-\theta)^{n-r} = \alpha, \tag{9.9a}$$

and an upper limit for θ is the solution to the equation

$$\sum_{r=0}^{x} \binom{n}{r} \theta^r (1-\theta)^{n-r} = \alpha. \qquad (9.9b)$$

These solutions may be deduced from tables of cumulative binomial probabilities or, more simply, read from a graph of confidence bounds, supplied in statistical tables such as Neave (1985), though with very small θ and large n, reading with any accuracy is difficult. One-sided intervals may be constructed in a similar way.

Example 9.15: Cracks in welded joints are assumed to be detected with the same probability, $1 - \theta$, independently of each other. Suppose it is desirable that the method chosen to detect cracks is at least 90% reliable. A detection method might be tested by selecting n cracks from a known population of cracks; to be satisfactory the method must detect all n cracks. The question of interest is how large n must be for 95% confidence that $1 - \theta$ is at least 0.9.

In requiring reliability of crack detection to be a minimum of 0.9 at each crack, the upper limit on θ is 0.1. We substitute this value into (9.9b) with α at its minimum of 0.05.

$$\sum_{r=0}^{0} \binom{n}{r} (0.1)^r (0.9)^{n-r} < 0.05$$

$$(0.9)^n < 0.05$$

$$n > \frac{\log 0.05}{\log 0.9} = 28.4$$

(Note the reversal of the inequality when dividing by a negative value.) So the sample size must be at least 29 and all 29 cracks must be detected in order to be 95% confident that θ is at least 0.9.

Use of the Poisson distribution

Where θ (or $1 - \theta$) is very small and n is large, an alternative formulation via the Poisson distribution may be considered. It can be shown that under these conditions the Poisson distribution is a good approximation to the binomial distribution. A derivation is given in Appendix A1 of Metcalfe (1994), for example. The occurrence of defectives may be considered a Poisson process with parameter μ representing the average number of defectives (Section 2.2).

A 100 $(1 - \alpha)$% upper limit for μ is found by solving $P(X \leq x) = \alpha$, where x is the number of observed defectives, that is,

$$\sum_{r=0}^{x} e^{-\mu} \mu^r / r! = \alpha. \tag{9.10}$$

Alternatively a chart of Poisson cumulative probabilities may be used, as in Neave (1985, p. 17).

Example 9.16: Newly manufactured 'threads' may be unwrapped to form a continuous length which is assumed to have random defects due to the machining process. In one particular heavy engineering application detectable defects are considered to be approximately 3 mm long and a metre of thread may be considered to be approximately 33 sub-lengths each of which may or may not contain a defect. Each sub-length thus forms an independent trial with 0 or 1 defect and the number of defects observed in a set of sub-lengths therefore can be considered to have a binomial distribution. It is anticipated that the number of defects found in each thread will be extremely small, possibly one at most.

If n metres are inspected there are approximately $33n$ independent trials. If zero defects are found, we can place an upper limit on the probability of a defect in any 3 mm sub-length. (The lower limit is obviously zero.)

If θ is the probability that a sub-length is defective, the probability that no defectives are found in $33n$ sub-lengths is $(1 - \theta)^{33n}$. An upper 95% limit for θ is given by solving (9.9b),

$$(1 - \theta)^{33n} = 0.05,$$

giving

$$\theta_U = 1 - (0.05)^{1/33n}.$$

The following are solutions for various values of n.

n (metres)	θ_U	n (metres)	θ_U
1	0.08678	10	0.00904
2	0.04438	20	0.00453
5	0.01799	50	0.00181

If one defect is found, θ_U is given by solving (9.9b) with $x = 1$,

$$(1 - \theta)^{33n} + 33n\theta(1 - \theta)^{33n-1} = 0.05.$$

Reference to binomial tables or charts is not easy because the value of θ is small and $33n$ is large. The equation may, however, be solved iteratively, using (say) Newton–Raphson or via computer algebra. If 10 metres of thread are inspected $\theta_U = 0.01429$.

The alternative formulation via the Poisson distribution may be considered, given that a sub-length is very small in relation to the length inspected. The occurrence of defects may be considered a Poisson process with parameter μ representing the expected number of defects in the inspected length. This can be converted to a rate of occurrence of defects per metre, say.

A 95% upper limit for μ is found by solving (9.10) with $x = 1$ and $\alpha = 0.05$:

$$\sum_{r=0}^{1} \frac{e^{-\mu}\mu^r}{r!} = 0.05,$$

giving

$$e^{-\mu}(1 + \mu) = 0.05.$$

Again, solving by iteration, we find $\mu_U = 4.744$. This is the upper limit on defects over the length inspected, so converting to defects per sub-length, we have $\theta_U = 4.744/(33 \times 10) = 0.01438$.

Using Poisson cumulative probability charts, the value of p is 0.95 and the value of x is 2. The upper limit for μ is therefore approximately 4.7. The Poisson approach works well here because the number of 'units' inspected is large. To use the chart for a two-sided interval, say at 90%, then p would be taken at 0.05 with $x = 1$ and p at 0.95 with $x = 2$.

Finite-population sampling

Sampling without replacement (as opposed to with replacement) is the general rule in life testing. This does alter slightly at each trial the probability that a defective is chosen, but is a negligible effect if the population is very large or the sample proportion fairly small, say less than 10%. In Example 9.16 the population is effectively infinite because the manufacturing process is continuous. To take another example, a typical North Sea exploration platform might have 300 or so primary joints. Scheduled inspections would involve detailed examination of a random sample of joints, probably around 10% of the total, so the following analysis may need to be considered.

If a sample of size n is taken from a population of size N having $K = N\theta$ defectives, then the probability that the sample contains x defectives is

$$P(X = x) = \frac{\binom{K}{x}\binom{N-K}{n-x}}{\binom{N}{n}},$$

where

$$\binom{i}{j} = \frac{i!}{(i-j)!j!}.$$

This formula defines the *hypergeometric distribution*. As in the binomial case, the mean of $P = X/n$ is θ and p is still an unbiased estimate of θ, but the variability in P is reduced by a factor known as the finite-population correction factor (f.p.c.f.), given by

$$\text{f.p.c.f.} = \frac{N-n}{N-1}.$$

(Some texts approximate this to $1 - n/N$.) The standard deviation of P is scaled by the square root of the f.p.c.f. and results in confidence intervals for θ which are narrower than those in the case where N is infinitely large. The general wisdom is that this correction factor should be applied when $n/N > 0.1$. As examples,

$$N = 100, n = 20, \sqrt{\text{f.p.c.f.}} = 0.899,$$
$$N = 200, n = 30, \sqrt{\text{f.p.c.f.}} = 0.924,$$
$$N = 300, n = 30, \sqrt{\text{f.p.c.f.}} = 0.950.$$

To obtain confidence intervals for θ, similar equations to (9.9a) and (9.9b) need to be solved with hypergeometric probabilities used instead of binomial or Poisson. Extensive tables and graphs of bounds are not readily available for the hypergeometric distribution. However, use of the binomial intervals will only err on the conservative side and therefore often provides an acceptable alternative. The shorter intervals based on the hypergeometric distribution may be obtained approximately by multiplying the intervals between p and θ_L and p and θ_U by the square root of the f.p.c.f.

It should be noted that the confidence intervals described above are not in general symmetric about p. The skewness becomes more marked as θ approaches 0 or 1. In the applications envisaged here θ is expected to be very small and therefore approximate methods based on the normal distribution are not in general appropriate.

9.9   Non-parametric estimation and confidence intervals for $R(t)$

Reliability may be estimated non-parametrically (that is, with no model assumed) by the empirical reliability function. The reliability at a particular value t^* may be of interest. The value of $R(t^*)$ lies between 0 and 1 and in n independent trials we would expect $nR(t^*)$ lifetimes to exceed t^*. We can think of $R(t^*)$ as corresponding to $(1 - \theta)$ in the terminology of Section 9.8. The number of lives greater than t^* is a binomial variable, X, with mean $E(X) = nR(t^*)$ and variance $V(X) = nR(t^*)[1 - R(t^*)]$.

We are interested in $\hat{R}(t^*)$, our estimate for $R(t^*)$, which we may take to be of the form $(X - c)/n$. We have that

$$V\left(\frac{X-c}{n}\right) = V\left(\frac{X}{n}\right) = \frac{V(X)}{n^2} = \frac{\hat{R}(t^*)[1 - \hat{R}(t^*)]}{n},$$

so the standard error of $\hat{R}(t^*)$ is

$$\left\{\frac{\hat{R}(t^*)[1 - \hat{R}(t^*)]}{n}\right\}^{1/2}.$$

Given a 'large' value for n, we can take $\hat{R}(t^*)$ to be approximately normally distributed and a $100(1 - \alpha)\%$ confidence interval for $\hat{R}(t^*)$ is given by

$$\hat{R}(t^*) \pm z_{\alpha/2} \text{s.e.}[\hat{R}(t^*)],\tag{9.11}$$

where $z_{\alpha/2}$ is the $100(1 - \alpha/2)\%$ point of the standard normal distribution.

Now consider the case where there are right-censored observations and the reliability function is estimated using the product-limit or Kaplan–Meier estimator (see Section 3.4). Given that there are failures at times $t_1 < t_2 < t_3 < \ldots < t_k$, d_j the number of failures at time t_j and n_j the number of items at risk at time t_j, then

$$\hat{R}(t) = \prod_{j:\, t_j < t} \left(1 - \frac{d_j}{n_j}\right).$$

Greenwood (1926) gives the following formula for the standard error of $\hat{R}(t^*)$,

$$\hat{R}(t^*)\left\{\sum_{j:\, t_j < t^*} \frac{d_j}{n_j(n_j - d_j)}\right\}^{1/2}.$$

Approximate confidence intervals may be obtained by using this expression in (9.11).

Example 9.17: The following data concern times to failure observed during testing of 25 of a particular engine component:

0.44 2.41 3.07 3.08 3.14 3.20 3.92
4.29 4.51 4.98 5.12 5.59 5.85 5.96 6.01.

The test was terminated at the time of the 15th failure.

Suppose that we are interested in a lower estimate for the reliability at time $t = 5$. The Kaplan–Meier estimate of the reliability function, shown below, yields an estimate for $R(5)$ of approximately 0.6. An estimate of the standard error of $\hat{R}(5)$ is best given by the nearest observed failure time, 4.98.

t	n_j	d_j	$\hat{R}$	$\dfrac{d_j}{n_j(n_j - d_j)}$	$\sqrt{\sum \dfrac{d_j}{n_j(n_j - d_j)}}$
0.44	25	1	0.96	0.001667	0.040825
2.41	24	1	0.92	0.001812	0.058977
3.07	23	1	0.88	0.001976	0.073855
3.08	22	1	0.84	0.002165	0.087287
3.14	21	1	0.80	0.002381	0.100000
3.20	20	1	0.76	0.002632	0.112390
3.92	19	1	0.72	0.002924	0.124722
4.29	18	1	0.68	0.003268	0.137199
4.51	17	1	0.64	0.003677	0.150000
4.98	16	1	0.60	0.004167	0.163299
5.12	15	1	0.56	0.004762	0.177281

A 95% lower confidence limit for $R(5)$ is then given by

$$\hat{R}(5)\left[1 - 1.645\left\{\sum_{j:\, t_j < 5} \frac{d_j}{n_j(n_j - d_j)}\right\}^{1/2}\right] = 0.6[1 - 1.645 \times 0.1633] = 0.439$$

It is important to note, in this non-parametric context, that reliability estimates can only be made within the range of the uncensored data.

9.10 Estimating system reliability from subsystem test data

Test data on the subsystems of a given system often become available at stages during the development of the system. This may be from life tests or 'one-shot' trials. These data can be used to construct both point and interval estimates of system reliability which can then be compared with the system specifications. This is especially useful when the assembled system is expensive to test and/or when testing results in destruction of the system.

Suppose there are k subsystems with reliabilities $\phi_1, \phi_2, ..., \phi_k$. Then, in general, the system reliability is given by

$$\phi = f(\phi_1, \phi_2, ..., \phi_k)$$

for some function f.

If $\hat{\phi}_1$, $\hat{\phi}_2$, ..., $\hat{\phi}_k$ are the maximum likelihood estimates of the subsystem reliabilities then the corresponding maximum likelihood estimate of system reliability is $\hat{\phi} = f(\hat{\phi}_1, \hat{\phi}_2, ..., \hat{\phi}_k)$; this is known as the *invariance* property of maximum likelihood estimators. However, such estimates are biased even if the individual $\hat{\phi}_i$ are unbiased. Rosenblatt (1963) describes the more complicated minimum variance unbiased estimation.

Usually of principal interest is interval estimation, particularly the determination of lower confidence limits, for system reliability. The problem is in general difficult but there are a number of approximate methods for well-defined special cases. Use of the asymptotic normality of maximum likelihood estimators and the inversion of the likelihood ratio test are two very general methods, but both tend to give optimistic results for small test sample sizes.

Series systems and binomial subsystem test data

In Section 9.8 a single binomial sample with sample size n and number of failures x was considered, and confidence limits for the unit failure probability, θ, were derived. Now consider a series system with $k = 2$ subsystems. The reliability of the system is $\phi = \phi_1\phi_2$, where $\phi_i = 1 - \theta_i$, and a lower limit for ϕ would be of interest. Let the binomial subsystem test results be (n_1, x_1) and (n_2, x_2), where $\{x_i\}$ are the numbers of subsystem failures.

The formula analogous to (9.9b) is

$$\sum_{r_1, r_2 \le x_1, x_2} \prod_{i=1}^{n_i} \binom{n_i}{r_i}(1 - \phi_i)^{r_i}\phi_i^{n_i - r_i} = \alpha. \qquad (9.12)$$

This equation is satisfied by a range of values of $\phi = \phi_1\phi_2$. When all subsystem sample sizes are equal and $k \le 3$, computation is not too difficult and there are tables of lower limits for ϕ for a range of sample sizes and selected values of α; see Lipow and Riley (1960). When subsystem sample sizes are unequal and $k > 2$, computation of the lower limit becomes difficult; see Winterbottom (1974). The following is a useful approximation.

Lloyd and Lipow (1962) refer to the Lindstrom and Madden method, which is normally described in terms of the number of successes in each sample, so let $s_i = 1 - x_i$. We define n to be the smallest sample size and calculate

$$s = n \prod_{i=1}^{k} \frac{s_i}{n_i}.$$

Then s is treated as the number of successes in a single binomial sample of size n. The value of s may not be an integer so use of tables of cumulative binomial probabilities requires interpolation, which may be done linearly. The method gives a result which errs on the conservative side, but gives the same as (9.12) when s is an integer.

Example 9.18: Suppose there are three subsystems in series and the individual test results are $n_1 = n_2 = n_3 = 20$; $s_1 = 17$, $s_2 = 19$, $s_3 = 20$. Then $n = 20$ and $s = 20 \times 17/20 \times 19/20 \times 20/20 = 16.15$. From tables,

n	s	lower 90% limit
20	17	0.696
20	16	0.639

Linear interpolation gives 0.648. The exact 90% limit is 0.660.

Example 9.19: Now suppose the test samples to be of different sizes, $n_1 = 20$, $n_2 = 24$, $n_3 = 30$; $s_1 = 19$, $s_2 = 22$, $s_3 = 28$. Then $n = 20$ and $s = 20 \times 19/20 \times 22/24 \times 28/30 = 16.256$. Again interpolating between the limits above, the lower 90% limit in this case is 0.654.

Series systems and exponential times to failure

The following is a method due to Lieberman and Ross (1971) which examines the behaviour of the subsystems over a common time scale and makes use of the exact theory of Section 9.3. The method may be illustrated with just two subsystems in series, without any loss of generality. The times to failure for the two subsytems, $T_{(1, i)}$ and $T_{(2, j)}$, will be taken to be exponential with parameters λ_1 and λ_2. Suppose that testing for both subsystems is by Type II censoring with sample sizes n_1, n_2 and numbers of failures r_1, r_2, respectively. Then we have for subsystem 1,

$$U_i = (n_1 + 1 - i) \, (T_{(1, i)} - T_{(1, i-1)}) \, , \, i = 1, 2, ..., r_1$$

and for subsystem 2,

$$V_j = (n_2 + 1 - j)(T_{(2, j)} - T_{(2, j-1)}), \, j = 1, 2, \ldots, r_2.$$

This is the same decomposition of total time on test as described in Section 9.3. It then follows that the U_i and V_j are independent exponential variates with failure rates λ_1 and λ_2, respectively. Now consider the combined process, illustrated in Figure 9.4, noting that the time scale is accumulated total time on test. As long as both lines are still in operation the combined line is a Poisson process with rate $\lambda = \lambda_1 + \lambda_2$. The $\{W_k\}$ are inter-failure times for the combined process and $2\lambda W = 2\lambda(W_1 + W_2 + \ldots + W_r^*)$ has a $\chi^2_{(2r*)}$ distribution. An upper $100(1 - \alpha)\%$ confidence limit for λ is

$$\lambda_U = \chi^2_{(2r*; \, 1 - \alpha)}/(2W).$$

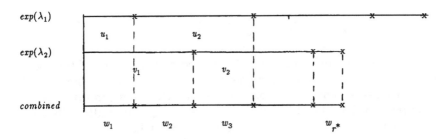

Figure 9.4. Combined Poisson process

Let the unit of time be system 'mission' time. Then the series system reliability is $\exp(-\lambda)$ and a corresponding lower limit for system reliability is $\exp(-\lambda_U)$.

Example 9.20: Consider the following observations for two subsystems, noting that $n_i - r_i$ is the number of right-censored observations for the ith subsystem: $n_1 = 6$, $r_1 = 3$, $t_{(1, 1)} = 7$, $t_{(1, 2)} = 18$, $t_{(1, 3)} = 27$; $n_2 = 10$, $r_2 = 5$, $t_{(2, 1)} = 4$, $t_{(2, 2)} = 6$, $t_{(2, 3)} = 10$, $t_{(2, 4)} = 20$, $t_{(2, 5)} = 41$. Figure 9.5 illustrates the calculation of the u_i and v_i. At total time on test equal to 133, the uncensored observations for subsystem 1 run out and therefore observation of the combined Poisson process stops. The number of observations for the combined process, r^*, is therefore $3 + 3 = 6$ and $w = 133$. The upper 90% confidence limit for λ is $\lambda_U = \chi^2_{(12; \, 0.9)}/(2w) = 18.55/266 = 0.0697$. Taking unit time to be the length of a mission, the corresponding lower 90% confidence limit for mission reliability is $\exp(-0.0697) = 0.933$.

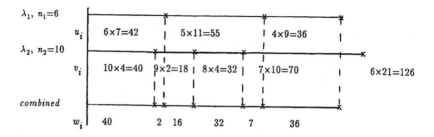

Figure 9.5. Combined Poisson process for Example 9.20

This method used as uncensored observations 6 of the 8 total such observations in the above example. If too many of the original observations are 'lost' then the method will lose some of its power.

General structures and maximum likelihood

As previously stated, maximum likelihood and inversion of the likelihood ratio test are very general and efficient procedures when sample sizes are large. The likelihood ratio approach is marginally superior to maximum likelihood but is, in general, computationally difficult.

Consider the maximum likelihood approach. Using the Taylor approximation of (7.10), the approximate large-sample variance of the maximum likelihood estimator of ϕ is

$$V(\hat{\phi}) = \sum_{i=1}^{k} \left(\frac{\partial \phi}{\partial \phi_i}\right)^2 V(\hat{\phi}_i).$$

For binomial data $V(\hat{\phi}_i) = \phi_i(1 - \phi_i)/n_i$, where the ϕ_i are the subsystem reliabilities.

Example 9.19 (continued): Here $\phi = \phi_1\phi_2\phi_3$ and $\dfrac{\partial \phi}{\partial \phi_i} = \phi_j\phi_k, i \neq j \neq k$. So,

$$V(\hat{\phi}) = (\phi_1\phi_2)^2\frac{\phi_3(1 - \phi_3)}{n_3} + (\phi_3\phi_2)^2\frac{\phi_1(1 - \phi_1)}{n_1} + (\phi_1\phi_3)^2\frac{\phi_2(1 - \phi_2)}{n_2}$$

$$= \phi^2\sum_{i=1}^{3}\frac{(1 - \phi_1)}{n_i\phi_i}.$$

All parameters are replaced by their maximum likelihood estimates. The $\hat{\phi}_i$ are s_i/n_i and $\hat{\phi} = \hat{\phi}_1\hat{\phi}_2\hat{\phi}_3$. An approximate lower confidence limit for ϕ is given by $\hat{\phi} - z_c\sqrt{V(\hat{\phi})}$, where z_c is the appropriate percentage point from $N(0, 1)$. For the data of this example, a lower 90% limit for ϕ is

$$0.8128 - 1.2816 \times 0.8128 \times \sqrt{0.0026316 + 0.0037879 + 0.0023810} = 0.715 .$$

The result for this example is optimistic. The convergence to normality is very dependent on the smallest test sample size. We can improve the above result by incorporating a correction based on the theory of asymptotic expansions, given in Stuart and Ord (1994). Let

$$\sum_{i=1}^{k} \frac{(1 - \phi_i)}{n_i\phi_i} = \frac{\sigma^2}{n} .$$

Then $V(\hat{\phi}) = \hat{\phi}^2\sigma^2/n$. The improved lower limit is given by

$$\hat{\phi}\left[1 - z_c\frac{\sigma}{\sqrt{n}} - \frac{1}{6\sigma^2}\left\{ [z_c^2 + 1]\left[n\sum_{i=1}^{k} \frac{(1 - \hat{\phi}_i^2)}{n_i^2\hat{\phi}_i^2} - 3\sigma^4 \right] + n\sum_{i=1}^{k} \frac{(1 - \hat{\phi}_i)(1 - 2\hat{\phi}_i)}{n_i^2\hat{\phi}_i^2} \right\} \right] .$$

A further improvement is a 'continuity correction' given by replacing $\hat{\phi}$ by $\hat{\phi} - \phi'/2n'$, where ϕ' is the maximum likelihood estimate of system reliability excluding the subsystem for which the test sample size is smallest and n' is the corresponding sample size. Applying both of these improvements to Example 9.19 gives a lower 90% limit for ϕ of 0.682.

A development of the use of asymptotic expansions is given in Winterbottom (1980).

Example 9.21: Now suppose that the two subsystems tested in Example 9.20 are assembled in parallel. Assuming that each has an exponential lifetime distribution, the reliability of the system is given by (6.5),

$$R(t) = e^{-\lambda_1 t} + e^{-\lambda_2 t} - e^{-(\lambda_1 + \lambda_2)t} ,$$

so that

$$\frac{\partial R}{\partial \lambda_i} = -te^{-\lambda_i t} + te^{-(\lambda_1 + \lambda_2)t} .$$

From (4.4) and Example 4.5, $\hat{\lambda}_i = r_i/T_i$ and $V(\hat{\lambda}_i) = \hat{\lambda}_i^2/r_i$. $T_1 = 133$ and $T_2 = 286$, so $\hat{\lambda}_1 = 3/133$ and $\hat{\lambda}_2 = 5/286$.

Applying (7.6),

$$V(\hat{R}) \approx \left(\frac{\partial R}{\partial \lambda_1}\right)^2 V(\hat{\lambda}_1) + \left(\frac{\partial R}{\partial \lambda_2}\right)^2 V(\hat{\lambda}_2)$$

$$= t^2 e^{-2\hat{\lambda}_1 t}\left(e^{-\hat{\lambda}_2 t} - 1\right)^2 \frac{\hat{\lambda}_1^2}{r_1} + t^2 e^{-2\hat{\lambda}_2 t}\left(e^{-\hat{\lambda}_1 t} - 1\right)^2 \frac{\hat{\lambda}_2^2}{r_2}.$$

Substituting for r_i and the estimated λ_i yields

$$V(\hat{R}) \approx t^2 [e^{-0.0451t}(e^{-0.0175t} - 1)^2(1.696 \times 10^{-4})$$
$$+ e^{-0.03497t}(e^{-0.02256t} - 1)^2(6.113 \times 10^{-5})]$$

Let mission length be unit time. Then an approximate lower 95% confidence limit for the mission reliability is

$$\hat{R}(1) - 1.645\sqrt{e^{-0.0451}(e^{-0.0175} - 1)^2(1.696 \times 10^{-4}) + e^{-0.03497}(e^{-0.02256} - 1)^2(6.113 \times 10^{-5})}$$
$$= 0.999613 - 1.645(2.796 \times 10^{-4}) = 0.999153.$$

Winterbottom (1984) is a review paper on the topic discussed in this section.

9.11   Accelerated testing

Accelerated life testing concerns the collection of lifetime data more quickly than would be the case in the normal use of components. For example, an electric toaster may generally only be used for half an hour a day. It is possible to simulate years of use by repeatedly using the toaster over a few weeks, making sure that at the start of each use the toaster is in 'normal' starting condition, for example, cool. Some tests may not need this period of 'recovery', such as repeated use of a door catch.

Other accelerating variables pose different problems. Often, in order to induce failure in a short time, it may be necessary to increase the severity of a condition such as temperature, load or vibration. The results of any of these tests have to be extrapolated back to the conditions of normal use, and care is needed in choosing the model on which to base this. One potential problem in accelerating a test is the possibility of introducing a mode of failure which would not normally be observed. It is also important that all failure modes occurring at low levels of the accelerating variables are seen.

Where the accelerating variable is, say, temperature or stress, knowledge of the physics of failure of the device is important in assessing an appropriate model to use. A tractable framework is one where it is assumed that the type of lifetime distribution does not vary with the accelerating variable, but the parameters of the distribution are functions of the level of the accelerating variable. For example, an exponential lifetime model with rate λ might be considered suitable with $\lambda = \lambda(s)$, where s is the accelerating variable. Forms which might be considered for λ include a power law,

$$\lambda(s) = Cs^k,$$

or the Arrhenius rate model,

$$\lambda(s) = C\exp(-k/s),$$

where C and k require estimation. Even under high levels of s, there may be censored observations, which result in the use predominantly of maximum likelihood techniques.

Where the form of an underlying model cannot be assumed or relationships of parameters with accelerating variables are uncertain, there is the option of applying non-parametric methods. However, these depend on some data at normal use conditions, which may be a significant drawback.

The execution and analysis of accelerated life tests is in general a complex area. A comprehensive text on the subject is Nelson (1993).

CHAPTER 10

Advanced Models

10.1   Covariates

It is frequently the case that time to failure is dependent on other random variables – characteristics which are perhaps subject to natural variation, but may also be under a certain amount of control. These *explanatory variables* or *covariates* influence the lifetime model through the reliability function, and thus by implication through the hazard function. Common to such models is the notion of a *baseline* reliability function which corresponds to the lifetime behaviour for some standard or initializing condition.

The covariates may be represented by a vector $\mathbf{Z} = \{Z_1, Z_2, ..., Z_n\}$ and may be continuous or discrete. This leads to a reliability function which is conditional on the vector $\mathbf{Z}$:

$$R(t; \mathbf{z}) = P(T \geq t \mid \mathbf{Z} = \mathbf{z}).$$

The covariates may be time-dependent, say $z = z(t)$. They may be influenced over time by the unit itself (for example, level of corrosion), or may not be directly involved in the failure process (for example, applied stress).

The general area of *regression modelling* covers a range of techniques appropriate to data for which there are covariates. In reliability there is widespread applicability of models based on the Weibull distribution. A definitive discussion of Weibull regression models in reliability is given in Smith (1991) and in Chapter 4 of Crowder *et al.* (1991).

There are essentially two principal objectives in the modelling process: to establish the form of the 'baseline' model, and to determine how the model parameters vary with the covariates.

10.2   Proportional hazards models

In the proportional hazards model, first proposed by Cox (1972), the combined effect of the z variables is to scale the hazard function up or down. The hazard function satisfies

$$h(t;\mathbf{z}) = h_0(t)\ g(\mathbf{z}) \tag{10.1}$$

where $h_0(t)$ is the baseline hazard function. Since

$$\frac{h(t;\mathbf{z}_1)}{h(t;\mathbf{z}_2)} = \frac{g(\mathbf{z}_1)}{g(\mathbf{z}_2)}$$

is independent of t, the hazards at different z values are in constant proportion over time. The function $g(\mathbf{z})$ may take a variety of forms involving functions of z_i, such as $\log z_i$.

It can be shown from (10.1) that

$$\log R(t;\ \mathbf{z}) = g(\mathbf{z})\log R_0(t)$$

and hence

$$R(t;\ \mathbf{z}) = [R_0(t)]^{g(\mathbf{z})}.$$

Simple graphical analysis of data follows by plotting $\log[-\log \hat{R}(t;\ \mathbf{z})]$ for different $\mathbf{z}$ against some function of t, typically t or $\log t$. Where fixing of the covariate at specified levels has not been possible, data may need to be grouped by putting similar values of z together. If the proportional hazards assumption holds, the plots should be similar curves shifted in the y-axis direction. In particular, if plotting against $\log t$ and the curves are approximately straight lines, then the Weibull model holds.

The weakest-link relationship is a special case of proportional hazards. If z is some size covariate, length say, the weakest-link model is given by

$$R_z(t) = [R_1(t)]^z.$$

Here $R_1(t)$ may be thought of as the baseline reliability function and $g(z) = z$ is a one-dimensional linear function. If, in addition, the proportional hazards plots are linear, then the Weibull model is appropriate and may be the basis for prediction at further values of z, as demonstrated in Example 2.2. It is seen in the latter that one of the model parameters is a function of the size covariate, n. Given baseline scale and shape parameters α and β, $\alpha(n) = \alpha n^{-1/\beta}$.

Example 10.1: Data from Wolstenholme (1995) were examined by a non-parametric method in Example 4.14 for conformance to the weakest-link model. The data refer to strength measurements of 24 carbon fibres coated in resin, where each had been subdivided into the four lengths, 5, 12, 30 and 75 millimetres, and then each length tested separately. There were, in addition, a small number of observations collected from further fibres where some sub-lengths had been damaged. The observations are all uncensored and plots for the complete data sets of $\log[-\log \hat{R}(t)]$ against $\log t$, where t is strength, are shown in Figure 10.1. We already know from Example 4.14 that there is some doubt as to whether the weakest-link model holds here. It can be seen that proportional hazards holds approximately, that is, the plots are shifted similar versions of each other, and further that, due to the linearity, Weibull models fit each data set, with a slight departure in the 5mm case. It is these slightly unusual latter observations that cause the doubt in Example 4.14.

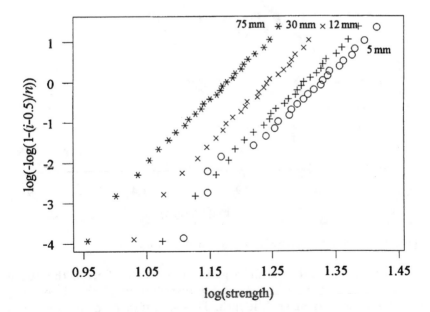

Figure 10.1. Proportional hazards plots for coated carbon

Example 10.2: Also in Wolstenholme (1995) is an analysis of a much larger data set of similar construction, but this time for uncoated carbon fibres. Overall there were 137, 139, 132 and 133 observations at the lengths 5, 12, 30 and 75 mm, respectively. The data for the 5 mm length are shown in Table 4.2 and assessed for goodness of fit to a Weibull distribution via the Pearson test in Example 4.13. Plots of $\log[-\log \hat{R}(t)]$ against $\log t$ are shown in Figure 10.2. From a materials science point of view the difference between this and Figure 10.1 is striking. The considerable increase in strength and the much reduced strength variability are the result of the resin coating. What is also clear is that the model for the strength is far from simple. First, the proportional hazards model does not apply and therefore neither does the weakest-link model. Second, the non-linearity of the plots indicates that none of the data sets can be said to have come from a Weibull distribution. Formal goodness-of-fit criteria, as applied in Example 4.8, reject the Weibull model in all cases.

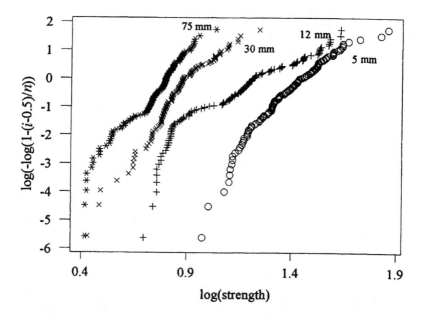

Figure 10.2. Proportional hazards plots for uncoated carbon

The experimental procedure giving rise to the data of the above examples ensured that the diameter distribution of the fibres concerned was similar at each length. This was important because it is known that like length, diameter also influences strength, being also a measure of 'size'. Figure 10.3 shows how strength varies with diameter for the 5 mm carbon data of Table 4.2. It could therefore be the case that a covariate vector, g(length, diameter), is a sensible proposition. Watson and Smith (1985) suggested one form of a fibre strength model which incorporated diameter into the Weibull scale parameter, though the data used did not give convincing results.

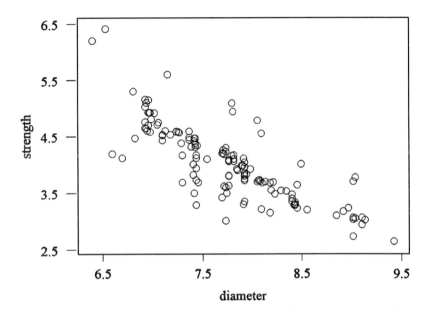

Figure 10.3. Strength as a function of diameter for the 5mm carbon data of Table 4.2

In the *linear proportional hazards model* it is assumed that the hazard function is of the form

$$h(t;\ \mathbf{z}) = h_0(t)\ \exp\{\mathbf{z}^{\mathrm{T}}\boldsymbol{\beta}\} = h_0(t)\ \exp\{\beta_1 z_1 + \beta_2 z_2 + \ldots + \beta_n z_n\}.$$

The ensuing analysis depends on whether we wish to assume or specify a particular form for the baseline hazard. A general starting assumption might be a power form for h_0; however, it is possible to estimate the coefficients $\{\beta_i\}$ without a given form for h_0 based on a type of likelihood function for $\boldsymbol{\beta}$.

Suppose a random sample of n units yields r uncensored lifetimes and $n - r$ right-censored lifetimes. Let the ordered lifetimes be $t_{(1)}$, $t_{(2)}$, ..., $t_{(r)}$, let C_i be the set of units at risk of failing just prior to $t_{(i)}$ and let $\mathbf{z}_{(i)}$ be the vector of covariates associated with the unit observed to fail at time $t_{(i)}$. The conditional probability that the unit corresponding to $z_{(i)}$ is the next to fail is

$$\frac{h(t_{(i)};\mathbf{z}_{(i)})}{\sum\limits_{k \in C_i} h(t_{(i)};\mathbf{z}_{(k)})} = \frac{\exp\{\mathbf{z}_{(i)}^{T}\,\boldsymbol{\beta}\}}{\sum\limits_{k \in C_i} \exp\{\mathbf{z}_{(k)}^{T}\,\boldsymbol{\beta}\}}.$$

The likelihood is the product of these terms over all failure times $t_{(i)}$. It is approximate when used for data sets containing ties, but provided the ratios d_i/n_i are small, where d_i is the number of units failing at time $t_{(i)}$ and n_i the number of units at risk at $t_{(i)}$, then the approximation is satisfactory.

Example 10.3: Consider a single covariate z used at l different levels and m distinct uncensored observations obtained at each z level. So we have $n = r = lm$. The log likelihood is

$$\log L(\beta) = \sum_{i=1}^{n} (\beta z_{(i)} - \log c_i),$$

where $c_i = \sum_{k \in C_i} \exp(\beta z_k)$, that is the $e^{\beta z}$ accumulated over all observations which were potential failures at $t_{(i)}$. Thus

$$\frac{\partial \log L(\beta)}{\partial \beta} = \sum_{i=1}^{n} z_{(i)} - \sum_{i=1}^{n} \frac{\sum\limits_{k \in C_i} z_k \exp(\beta z_k)}{c_i},$$

which may be solved numerically to find $\hat{\beta}$. The approximate standard error of $\hat{\beta}$ may be found in the usual way from the second derivative of $\log L(\beta)$.

A thorough exposition of proportional hazards models and their analysis can be found in Cox and Oakes (1984).

10.3   Accelerated life models

In accelerated life models, time is scaled by a function of the covariates:

$$R(t; \mathbf{z}) = R_0(g(\mathbf{z})t),$$

where R_0 is some baseline reliability function and $g(\mathbf{z})$ is a positive function of $\mathbf{z}$. The decrease in the survival probability is accelerated by a factor $g(\mathbf{z})$. The scaled lifetime $Y = g(\mathbf{z})T$ has reliability function R_0. Also,

$$\log Y = \log [g(\mathbf{z})] + \log T$$

and hence

$$\log T = \log Y - \log [g(\mathbf{z})].$$

Therefore the distribution of $\log T$ is the distribution of $\log Y$ subject to a location shift. By differentiation, it can be shown that the hazard function satisfies

$$h(t; \mathbf{z}) = g(\mathbf{z})h_0(g(\mathbf{z})t).$$

Example 10.4: Suppose $g(\mathbf{z}) = z^\beta$, for some $\beta > 0$ and $R_0(t) = e^{-t}$. Then $R(t; z) = \exp(-z^\beta t)$, so T is exponentially distributed with mean $z^{-\beta}$. Thus it can be seen that the Weibull model is both an accelerated life and a proportional hazards model.

10.4   Mixture models

It may be known that, for a particular type of unit, there are two or more distinct causes of failure. If, *for each observed failure*, the cause of failure can be identified, then the data may be partitioned into separate sets for each failure mode and under certain conditions a failure distribution fitted to each mode separately. However, a distinction must be drawn between the following cases:

(a)   Each unit has only one possible failure mode.
(b)   Each unit has k possible failure modes, independent of each other.
(c)   As (b), except certain modes of failure may be dependent on others.

In (a) the partitioning of the data is simple and each data set may be treated independently. Suppose a population of devices has a proportion p with lifetime reliability function $R_1(t)$ and proportion $1 - p$ with lifetime reliability function $R_2(t)$. Let T represent the lifetime of a random member of the population. The reliability function is $R(t) = P(T > t)$. Recalling probability rule (1.5),

$$P(T > t) = P(\text{type 1 device})P(T > t \mid \text{type 1 device}) \\ + P(\text{type 2 device})P(T > t \mid \text{type 2 device}),$$

so that

$$R(t) = pR_1(t) + (1 - p)R_2(t). \qquad (10.2)$$

Given n observations, of which n_1 are failures of type 1 units and $n_2 = n - n_1$ are failures of type 2 units, the model fitting techniques of Chapters 3 and 4 may be applied to each group of failure times. These observations are by definition uncensored, since the mode of failure is known. If the proportion p is unknown, then a simple estimate of p is n_1/n.

The form (10.2) is a *finite mixture model* and may be extended in a general way to accommodate k modes of failure.

Example 10.5: Let two types of component have exponential lifetimes with means θ_1 and θ_2, respectively. For a population of these components in which a proportion p are type 1, the reliability function for a random member of the population is

$$R(t) = pe^{-\lambda_1 t} + (1 - p)e^{-\lambda_2 t},$$

where $\lambda_1 = 1/\theta_1$ and $\lambda_2 = 1/\theta_2$. The probability density function is given by

$$f(t) = -R'(t) = p\lambda_1 e^{-\lambda_1 t} + (1 - p)\lambda_2 e^{-\lambda_2 t}. \qquad (10.3)$$

The hazard function is given by

$$\begin{aligned} h(t) = \frac{f(t)}{R(t)} &= \frac{p\lambda_1 e^{-t(\lambda_1 - \lambda_2)} + \lambda_2(1 - p)}{pe^{-t(\lambda_1 - \lambda_2)} + 1 - p} \\ &= \frac{\lambda_1 pe^{-t(\lambda_1 - \lambda_2)} + \lambda_1(1 - p) - \lambda_1(1 - p) + \lambda_2(1 - p)}{pe^{-t(\lambda_1 - \lambda_2)} + 1 - p} \\ &= \lambda_1 + \frac{(\lambda_2 - \lambda_1)(1 - p)}{pe^{t(\lambda_2 - \lambda_1)} + 1 - p}. \end{aligned}$$

The mean lifetime is

$$\mu = \frac{p}{\lambda_1} + \frac{1-p}{\lambda_1\lambda_2}.$$

Figure 10.4 shows the mixture model (10.3) for $\lambda_1 = 1$, $\lambda_2 = 2$ and $p = 0.5$. There appears to be little to distinguish the model from an exponential, and an exponential model with the same mean (0.75) as the mixture is shown for comparison.

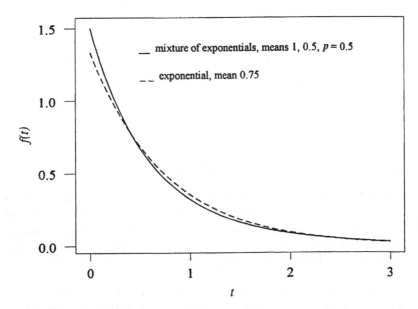

Figure 10.4. Exponential mixture models

A mixture model is most distinctive when it leads to a bi- or multimodal distribution. Figure 10.5 illustrates two cases of (10.3) where Weibull components are used, each having shape parameter 5 and scale parameter 1 and 2, respectively. The bimodal nature of the mixture distribution is a function of the value of p and the degree of overlap of the component distributions.

We can think of the unit lifetime in Example 10.5 being exponentially distributed with a rate parameter which takes the value λ_1 with probability p and the value λ_2 with probability $1 - p$. This leads to the idea of distribution parameters having themselves a probability distribution based on some characteristic of the population. The Pareto model of

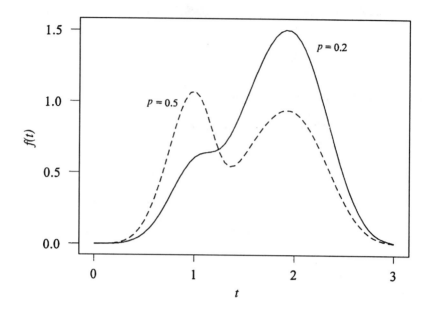

Figure 10.5. Finite mixtures of Weibull distributions

Section 2.8 was shown to arise as a mixture model in the case where the lifetime of a population is exponential, but the rate parameter over the population has a gamma distribution (called the *weighting* or *mixing* distribution). In reality, most populations will be mixtures in some sense, but explicitly modelling the mixing may not necessarily be worthwhile.

Example 10.6: This is a classic example of a mixture distribution. Suppose a process has failures which occur as a Poisson process with mean μ in an operating interval T. A failure is a serious failure if the failure detection and fail-safe facility fails to work, which happens with probability θ for any failure. It is fairly easy to deduce that the mean number of serious failures in an interval T is $\mu\theta$, but the probability distribution of the number of failures may also be of interest.

For a given number of failures N, the number of serious failures, X, has a binomial distribution with parameters N and θ. So the distribution of X, conditional on N, is given by

$$p_{X|N}(x; n) = \frac{n!}{x!(n-x)!}\theta^x(1-\theta)^{n-x}.$$

But N is a random variable with $p_N(n) = e^{-\mu}\mu^n/n!$. The unconditional distribution of X is a mixture model, an extension of (10.2) to an infinite number of terms:

$$p_X(x) = \sum_{n=x}^{\infty} p_N(n)p_{X|N}(x; n), \quad n \geq x,$$

$$= \sum_{n=x}^{\infty} \frac{n!}{x!(n-x)!}\theta^x(1-\theta)^{n-x}\frac{e^{-\mu}\mu^n}{n!}$$

$$= \frac{e^{-\mu}}{x!}\theta^x \sum_{n=x}^{\infty} \frac{(1-\theta)^{n-x}\mu^n}{(n-x)!}$$

$$= \frac{e^{-\mu}}{x!}\theta^x \sum_{k=0}^{\infty} \frac{(1-\theta)^k\mu^{k+x}}{k!}, \quad k = n - x$$

$$= \frac{e^{-\mu}}{x!}\theta^x\mu^x \sum_{k=0}^{\infty} \frac{[\mu(1-\theta)]^k}{k!}$$

$$= \frac{e^{-\mu}(\mu\theta)^x e^{\mu(1-\theta)}}{x!} \quad \text{(see the Appendix for the series expansion of e^x)}$$

$$= \frac{e^{-\mu\theta}(\mu\theta)^x}{x!}.$$

Hence the number of serious failures has a Poisson distribution with parameter $\mu\theta$. The context for this model is similar to that of Example 8.10.

10.5   Competing risks

Case (b) in Section 10.4 requires what is termed a *competing risks* model. A unit displays failure mode m at time t, conditional on all other failure modes having a time to failure greater than t, that is, the modes of failure 'compete' to be the one with the lowest failure time. Human beings are subject to many competing potential causes of death. As advances in medical science make it less likely that death is due to some causes, the incidence of other causes rises, purely because in the past fewer people lived to the ages where these other causes are more prevalent. Similarly, in reliability, action to deal with a certain cause of failure does not eradicate failure but lengthens the time until failure by another cause.

For the moment, the case where the different failure modes are independent will be considered. Let T represent the time to failure, and T_i the time to failure by mode i. Then

$$R(t) = P(T > t) = P([T_1 > t] \cap [T_2 > t] \cap ... \cap [T_k > t])$$
$$= P(T_1 > t)P(T_2 > t)...P(T_k > t), \text{ since all } T_i \text{ independent (see (1.4))},$$
$$= R_1(t)R(t_2)...R_k(t).$$

$$(10.4)$$

Fitting a lifetime model to each failure mode requires care, since each observation yields information about every possible failure mode, and necessarily much of the information is in censored form.

Let there be n units on test. Each yields an observation (t_i, m), the time and mode of failure, or a censored time where no failure is observed. The lifetime data for mode j consists of all t_i for which $m = j$, together with censored observations of the form $t > t_i$, where $m \neq j$. If there is no failure, the unit yields a censored observation for all modes. Each mode of failure then has n observations and now may be analysed separately. The analysis will give satisfactory results provided the proportion of censored data is not too high. In other words, it is desirable to have only a small number of distinct failure modes and a high proportion of uncensored observations. Information will also be limited if a failure mode does not manifest itself very often. If there are n units on test and k possible failure modes, there are k data sets of n censored or uncensored observations, but only a maximum of n uncensored observations in total.

Example 10.7: The data in Table 10.1 are from Hinds (1996) and concern failures of engines fitted to heavy duty vehicles. There are 63 observations. For each failure, the miles travelled to failure and the cause of failure are reported. The data are listed in miles to failure order purely for analytical convenience; no time or other order is implied. All miles to failure data are considered to be recorded from similar initial conditions. Five causes of failure are identified: the cooling system, dirt contamination, mechanical failure, ignition fault and fuel fault. The data are illustrated in Figure 10.6. In total the last two faults occurred only six times, so that inference about these failure types is severely limited. All engines are subject to all of the faults and therefore a competing risks model is appropriate.

miles	type	miles	type	miles	type	miles	type	miles	type
30	1	381	2	752	2	1486	4	3839	2
75	1	489	1	797	1	1551	1	4064	2
80	2	503	2	815	4	1831	1	4065	1
110	1	514	2	875	1	1875	2	4363	3
151	3	520	1	896	2	2188	5	4567	3
175	1	591	3	977	2	2254	1	4698	1
215	2	600	5	992	2	2701	3	4837	1
274	3	610	1	1000	2	2835	3	5023	2
284	4	658	2	1050	2	2914	2	5047	1
325	2	731	4	1122	3	2976	3	6197	2
340	2	739	1	1214	1	3480	3		
360	3	747	3	1231	1	3651	1		
374	2	748	3	1275	3	3672	2		

Table 10.1. Engine failure data

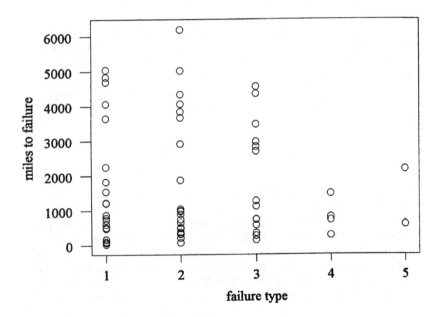

Figure 10.6. Engine failure data for Example 10.6

A highly important question is whether these causes of failure are independent. Hinds concluded from engineering knowledge and a correlation analysis that the three main causes of failure were effectively independent. On that assumption, we can apply (10.4) to give

$$R(t) \approx R_c(t) R_d(t) R_m(t)$$

where c, d, m refer to cooling, dirt and mechanical faults. The random variable T is mileage to failure.

There are 20 failures attributed to failure type 1, cooling system failure. Estimation of the reliability function $R_c(t)$ uses these 20 uncensored lifetimes together with the other 43 lifetimes, all treated as right-censored observations. Similarly, we may estimate $R_d(t)$ and $R_m(t)$. Hinds fits a Weibull model to each mode of failure and concludes via probability plotting and maximum likelihood methods that an exponential model provides an adequate model in each case. A likelihood ratio test, as demonstrated in Example 9.13, can be used to show whether or not a Weibull model has a significant advantage over an exponential distribution.

The notion of independent competing risks is analogous to a series system, discussed in Chapter 6. Equation (6.4) shows the form of the reliability function where components have exponential lifetime. For failure mode i, the maximum likelihood estimate of the rate parameter λ_i is, from (4.4), r_i/T, where T is total time on test and r_i the number of uncensored observations of failure type i. If λ is the rate parameter for the whole system, then

$$\hat{\lambda} = r/T = (r_1 + r_2 + r_3 + \ldots)/T = \hat{\lambda}_1 + \hat{\lambda}_2 + \hat{\lambda}_3 + \ldots .$$

From Table 10.1, the total time on test is 108081 miles and the number of uncensored observations for failure types 1 to 5 are 20, 23, 14, 4 and 2, respectively. The maximum likelihood estimates of the λ_i are then respectively 20/108081, 23/108081, 14/108081, 4/108081 and 2/108081. The standard errors of these estimates are high in the latter two cases, and little real inference can be made there. However, the data as a whole show an acceptable conformance to a Weibull model with shape parameter not significantly different from 1 [Figure 10.7]. The goodness-of-fit tests of Section 4.12 may be applied, taking the case where the Weibull shape parameter is fixed, here taken to be 1. The exponential rate parameter is estimated to be 63/108081 and the formulae (4.11) and (4.12) appropriately scaled yield 0.1408 and 0.834, respectively. Both are below the corresponding 5% points in Table 4.3 and hence the exponential model for $R(t)$ is accepted.

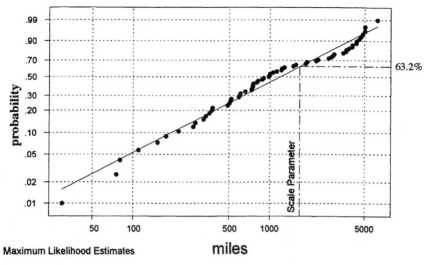

Figure 10.7. Weibull plot of Hinds' data

The important conclusion from Hinds' analysis was that since failure times tended to exhibit constant hazard the main route to increasing reliability was to address the causes of failure rather than look to maintenance as the principal factor. Dirt contamination was identified as the most frequently occurring failure type and it was estimated that fitting a more efficient air filtration system to the engines would have produced very high cost saving.

There is a cautionary note to competing risks analysis. *Identifiability* is a problem, discussed for example by Kalbfleisch and Prentice (1980), because for a given set of data it is always possible to fit equally well a model for independent risks and a model for non-independent risks. So it is essential that practical experience or exploratory analysis identifies a suitable form of model to be used.

10.6   Dependent failures

Dependency in the context of failure data can be presented in a variety of forms. Covariates enable failure of a single unit to be expressed as a function of some internal or external characteristic of the unit. In that sense failure is dependent on the level of that characteristic. In a repairable system, the existence of trend over time indicates that the inter-failure times are not independent identically distributed variates. It might be that they are independent but not identically distributed, or perhaps not independent either. There could be correlation between successive failure times, or the repair process itself may introduce some covariate.

Failure dependency is more commonly discussed in the context of systems and the possible interaction between component failures. The failure of one unit may produce overloading or overheating (say) of adjacent units whose failure is then hastened (see Section 10.7). Coincidental failures may be due not to dependence between units but to a common dependency on an external factor, such as power failure. This common cause failure has previously been described in Section 6.10. The common cause of failure acts in series with the components affected. Here, and in general, a unit may be subject to both independent failure and failure dependent on other units or common factors. The *beta factor* model, first introduced by Fleming (1975), partitions p, the probability of failure of a unit, into the probability of independent failure, $(1 - \beta)p$, and the probability of dependent failure, βp. The idea is simple, but to use the model effectively some subjective judgement is required.

Failure dependency within a series system, or equivalently a single unit under competing risks (case (c) in Section 10.4), is difficult to identify since only one mode of failure is observed and the residual effect on other modes of failure cannot usually be quantified. Hence the reliance on engineering judgement or experiments specifically designed to record relevant dependency indicators.

Dependency is discussed in detail in Chapter 8 of Ansell and Phillips (1994), and a review paper on the topic is Ansell and Walls (1992).

10.7   Load-sharing systems

A common type of failure dependency arises where a component of a system under a specified load may have its failure time altered by the failure of some of the components and the consequent redistribution of load. The strength of a system of n components sharing the applied load equally is the largest load that can be applied without every component failing.

Consider a simple system with two components and load per component x. Let $F(x)$ be the strength distribution function for individual components and $F_n(x)$ the strength distribution function for a system of n components. In the two-component system, when one component fails the load on the other component is $2x$. The system fails if one component has strength less than x and the other has strength between x and $2x$, or if both components have strength less than x. By summing the probabilities for these three mutually exclusive events, we can find $F_2(x)$:

$$F_2(x) = F(x)[F(2x) - F(x)] + [F(2x) - F(x)]F(x) + F(x)^2$$

$$= 2 \ F(x)F(2x) - F(x)^2.$$

By further extending what is effectively the application of the inclusion–exclusion formula, (6.1), a general expression for $F_n(x)$ may be obtained. However, the calculation becomes very complicated for large n.

One approximate approach centres on load per component, rather than total load on the system. It can be shown that the strength of the system is asymptotically normal, and expressions for the mean and variance can be formulated in terms of the maximum load per component. Analysis of this model largely originates from Daniels (1945).

When the load per surviving component is x, the number of surviving components is expected to be $n[1 - F(x)] = nR(x)$. The total load on the system is therefore $xnR(x)$, or $xR(x)$ per component. We shall assume that $xR(x)$ has a unique maximum at $x = x^*$. Various approximations have been derived for the distribution of x^*. A numerical study of such approximations was carried out by McCartney and Smith (1983). Barbour (1981) shows that x^* is approximately normally distributed with mean μ_* and variance σ_*^2 given by

$$\mu_* = M + 0.996n^{-2/3}B, \ \sigma_*^2 = n^{-1}[D - 0.317n^{-1/3}B^2] , \qquad (10.5)$$

where

$$M = x^*R(x^*), \ D = (x^*)^2F(x^*) \ R(x^*), \ B^3 = \frac{(x^*)^4[F(x^*)]^2}{2F'(x^*) + x^*F''(x^*)} .$$

Example 10.8: Suppose that component strength is Weibull distributed with shape parameter β. Without loss of generality, the scale parameter is set to 1. For a system of n such components the load per component will be denoted by y:

$$y = xR(x) = \exp[-x^\beta]$$

$$\frac{dy}{dx} = (-\beta x^\beta + 1)\exp[-x^\beta]$$

$$= 0 \text{ when } \beta x^\beta = 1.$$

Hence,

$$x^* = \left(\frac{1}{\beta}\right)^{1/\beta}$$

We find in this case that $M = \beta^{-1/\beta}\exp[-1/\beta]$, $D = M^2(\exp[1/\beta] - 1)$ and $B = M\beta^{-1/3}\exp[2/3\beta]$.

Using (10.5), approximate 95% confidence intervals for the strength of the system are given by

$$\mu_* \pm 1.96\sigma_*$$

Table 10.2 shows such confidence intervals for systems of 20 and 100 components and β given the values 3, 5 and 10.

β	$n = 20$	$n = 100$
3	[0.434, 0.675]	[0.460, 0.574]
5	[0.540, 0.754]	[0.561, 0.663]
10	[0.680, 0.854]	[0.693, 0.777]

Table 10.2. 95% confidence intervals for the reliability of load-sharing systems of n units, each unit with Weibull distributed lifetime, scale 1, shape β

10.8 Bayesian reliability

Bayesian methods incorporate the idea, met in Section 10.4, that distribution parameters may be treated as random variables. The results of an experiment are combined with beliefs based on earlier experimentation or engineering/management judgement. The more new data there are, the more those prior beliefs might be amended. The principle is based on Bayes' theorem (1.6). Given a set of data, $t = \{t_1, t_2, ..., t_n\}$, and an unknown distribution parameter θ,

$$P(\theta|t)P(t) = P(t|\theta)P(\theta).$$

The data are known, so that $P(t)$ may be regarded as a constant, hence

$$P(\theta|t) \propto P(t|\theta)P(\theta). \qquad (10.6)$$

The relationship (10.6) is valid for both discrete and continuous θ and t. In the continuous case P is a probability density function.

With an assumed model for the population from which the data are drawn, we can write down $P(t \mid \theta)$, since it is the likelihood function. $P(\theta)$ is an expression of our beliefs as to the likely value of θ. This is called the *prior distribution* of θ. This could be based on fitting a distribution to θ which has as *mode* (where the probability density is maximum) the value thought most likely for θ, and which matches some confidence interval estimate of θ. Substituting into (10.6) produces a form for $P(\theta|t)$, the *posterior distribution* of θ, that is the modified belief about θ, given the data observed. The posterior distribution of θ is made a properly defined probability density function by the inclusion of a *normalizing constant*, which is independent of θ. This ensures, in the continuous case, that $P(\theta|t)$ integrates to 1 over the range of θ, and, in the discrete case, that $P(\theta|t)$ sums to 1 over the range of θ. In some cases t will be replaced by a sample statistic, which is some function of t, when the sampling distribution of the statistic is straightforward.

If a second experiment is conducted, the posterior distribution of θ from the first experiment becomes the prior distribution of θ for the second, and so on. The same net result is obtained if all the data are combined and the original prior used. As more data are collected so the weight of the original beliefs is reduced.

It proves convenient to use, if possible, a *conjugate prior* distribution. This results in the posterior distribution of θ belonging to the same distribution family as the prior, but has in general different values for the distribution parameters. For example, a gamma prior for the rate parameter, λ, of the exponential distribution yields a gamma posterior distribution for λ. If the parameter to be estimated is θ in the binomial distribution $B(n, \theta)$ (Section 9.8), then the conjugate prior is the *beta distribution* which has probability density function

$$f(\theta) = B\theta^{a-1}(1-\theta)^{b-1},$$

where $B = \Gamma(a+b)/\Gamma(a)\Gamma(b)$. (See Section 2.3 for a definition of the gamma function, $\Gamma(a)$.) If a component has reliability θ and in a test of n components x failed, the posterior distribution of θ is beta with parameters $a^* = a + n - x$ and $b^* = b + x$.

Even where there is considerable uncertainty about the distribution parameter being estimated, a *vague prior* distribution may be used to good effect. The vaguest prior for the binomial parameter, θ, would be a uniform distribution on the interval [0, 1].

Example 10.9: Let t_1, t_2, ..., t_n be a set of observations from an exponential distribution with rate parameter, λ, considered to have a gamma distribution, parameters (α, β). In (10.6) the functions P are replaced by probability density functions:

$$f(\lambda|\mathbf{t}) \propto f(\mathbf{t}|\lambda) f(\lambda).$$

The function $f(\mathbf{t}|\lambda)$ is the likelihood

$$\prod_{i=1}^{n} \lambda e^{-\lambda t} = \lambda^n \exp\left(-\lambda \sum_{i=1}^{n} t_i\right) = \lambda^n e^{-\lambda T}.$$

The prior distribution of λ is

$$f(\lambda) = \frac{\beta^\alpha \lambda^{\alpha-1} e^{-\beta\lambda}}{\Gamma(\alpha)}.$$

We can ignore all terms which are not functions of λ, since these will all be absorbed into the normalizing constant. The terms involving λ are sufficient to identify the form of the posterior distribution of λ.

$$f(\lambda|\mathbf{t}) \propto \lambda^n e^{-\lambda T} \lambda^{\alpha-1} e^{-\beta\lambda} = \lambda^{\alpha+n-1} e^{-(\beta+T)\lambda}.$$

This shows that the posterior distribution of λ is gamma with parameters $(\alpha + n, \beta + T)$.

Suppose five observations are recorded, 0.03, 0.23, 1.03, 1.25, 2.83. So $n = 5$ and $T = 5.37$. Figure 10.8 shows the prior and posterior distributions of λ if the prior is gamma with parameters $\alpha = 2$ and $\beta = 3.2$.

Having acquired the posterior distribution of θ, point and interval estimates may be given for θ. The choice of estimate will depend on the purpose of the estimation and the potential 'loss' arising from the difference between the estimated and true value of θ. The idea is to choose an estimator for θ which in some sense minimizes the loss, or

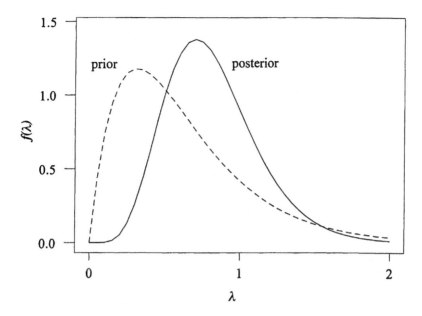

Figure 10.8. Gamma prior and posterior distribution for Example 10.9

Bayes' risk. Loss may be measured in a variety of ways and is not necessarily financial. If loss is proportional to the square of the difference between the true and estimated θ, then the optimal Bayes' estimate of θ is the posterior mean. If loss is proportional to the absolute difference, then the optimal estimate is the posterior median. In other circumstances, it might be appropriate to use the mode.

Example 10.10: Suppose a quality characteristic, X, of a product has been shown to be normally distributed with mean μ and variance σ^2. Suppose also that μ is considered to vary from one batch of the product to another as a normal variable with mean μ_0 and variance σ_0^2. We have

$$f(x|\mu) = \frac{1}{\sigma\sqrt{2\pi}}\exp\left[-\frac{1}{2}\left(\frac{x-\mu}{\sigma}\right)^2\right]$$

$$f(\mu) = \frac{1}{\sigma_0\sqrt{2\pi}}\exp\left[-\frac{1}{2}\left(\frac{\mu-\mu_0}{\sigma_0}\right)^2\right]$$

$$f(\mu|x) \propto f(x|\mu)f(\mu) \propto \exp\left[-\frac{1}{2}\left(\frac{x-\mu}{\sigma}\right)^2 - \frac{1}{2}\left(\frac{\mu-\mu_0}{\sigma_0}\right)^2\right].$$

By algebraic manipulation of the exponent, it can be shown that the posterior distribution of μ, $f(\mu \mid x)$, is a normal distribution with mean $(x\sigma_0^2 + \mu_0\sigma^2)/(\sigma_0^2 + \sigma^2)$. This expression will be the Bayes estimate for μ here, whether it is the posterior mean, median or mode that is required.

Suppose μ is to be estimated from a sample of 16 units selected at random from a new batch of the product. Let experience show that $\mu_0 = 200$ and $\sigma_0^2 = 25$ and that σ may be taken to be 40. The sample mean, $\bar{x}$, is a suitable statistic to use in the estimation of μ. Given that $x \sim N(\mu, \sigma^2)$, it follows that $\bar{x} \sim N(\mu, \sigma^2/n)$, where n is the sample size. Replacing x by $\bar{x}$ in the preceding theory, $f(\mu \mid \bar{x})$ is a normal distribution with mean

$$\frac{25\bar{x} + 200(40^2/16)}{25 + 40^2/16} = \frac{1}{5}\bar{x} + \frac{4}{5}200.$$

Thus past experience is given four times as much weight as the sample mean from the new batch. The weights are a function of the variances of x and μ.

Bayesian methods have attracted interest in reliability partly because they provide a way of incorporating engineering judgement. Design changes in products are often incremental and the characteristics of a new design are therefore not expected to be distant from those of the previous design. As systems become increasingly reliable, so the time and cost in collecting data become greater. Bayesian methods enable inference to be made based on a relatively small amount of data.

A wide-ranging introduction to Bayesian approaches in reliability analysis, with useful references, is given in Chapter 6 of Crowder *et al.* (1991).

10.9   Case studies

Example 10.11: Table 10.3 shows a famous set of motor failure data for buses, first given by Davis (1952), and which has been the source of much modelling interest since. Recent studies include Mudholkar *et al.* (1995) and Lindsey (1997). These data provide various modelling challenges and also illustrate some of the ways in which reliability data can be deficient.

The observations concern mileage between failures for 191 buses. Failure is a broken part and/or unacceptably impaired performance. There

Failure mileage (1000 miles) since last failure	observed frequency				
	1st	2nd	3rd	4th	5th
0-10	3	8	12	15	19
10-20	3	11	15	19	10
20-30	6	9	13	9	12
30-40	5	4	3	11	15
40-50	7	6	9	2	8
50-60	9	7	9	13	6
60-70	9	6	6	12	4
70-80	16	9	7	3	4
80-90	14	7	5	3	2
90-100	20	8	6	5	2
100-110	25	12	9	1	1
110-120	21	6	2	2	1
120-130	23	1	1	1	0
130-140	10	4	3	0	0
140-150	11	2	0	0	0
150-160	5	0	0	0	0
160-170	0	1	0	0	0
170-180	2	0	0	0	0
180-190	1	1	1	0	0
190-200	0	1	0	0	0
200-210	0	1	0	0	0
210-220	1	0	0	0	0
Total observations	191	104	101	96	84

Table 10.3. Bus motor failure data (from Davis, 1952)

are up to five failures recorded for each bus, and after the first failure motors have a mixture of new and used parts. The modelling focus has been on finding lifetime models for each set of inter-failure distances, making the assumption that the mileage to failure j is independent of the mileage to failure i. Each set of distances has a very different distribution from the others. The estimated distribution function for each data set is shown in Figure 10.9. In one sense there is a fairly consistent change in the distances to failure as failures are accumulated: the distances to failure tend to decrease over time. There is a clear difference in the nature of the distribution for first failure compared to subsequent failures, possibly for the reason already noted, namely that up to first failure all parts are new. Lindsey (1997) shows that the hazard function for the first failure climbs steeply after about

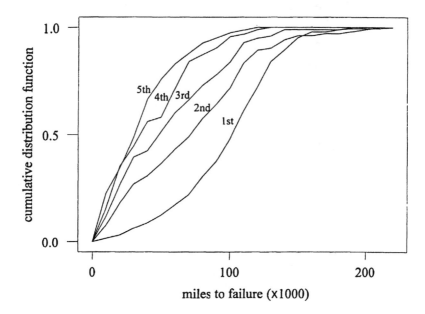

Figure 10.9. Empirical distribution functions for the bus data of Example 10.11

100000 miles, but for subsequent failures the hazard is quite close to linear after an initial risk which rises after each failure.

A common starting point with these data sets is to look at the fit of a Weibull model. Figure 10.10 shows Weibull plots for each data set. For every data set the fit is poor. There is evidence that over the second to fifth failures a degree of proportional hazards might apply, which is the theme of the modelling done by Lindsey (1997). A feature of most of the models that have been proposed is that there is little physical interpretation. This is undoubtedly linked to the deficiencies of the data set.

In Table 10.3 there are different numbers of failures in each column. There is no information as to why 107 buses eventually leave the data set, the most significant reduction in the size of data set occurring after the first failure. We really need to know whether a bus is retired at a failure due to the nature of the failure or for some other reason, or whether retirement is unassociated with failure. In the latter case there would be a censored contribution to the relevant data set. Such contributions could make a considerable difference to the failure mileage distribution. Example 3.3 illustrates a potential effect. Further, a full

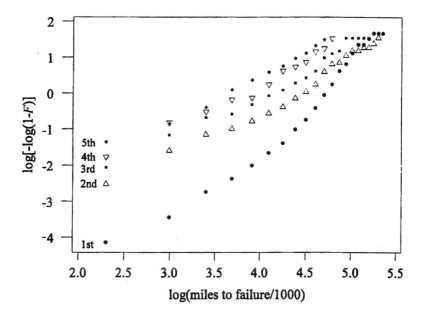

Figure 10.10. Weibull plots for the bus data of Example 10.11

history for each bus is required (along the lines of Example 5.2) in order to determine whether or not inter-failure mileages are correlated. The existence of trend over time may be examined, as in Section 5.3, but it is more important to establish whether the mileage between failure i and failure $i + 1$ for each bus can be treated as belonging to the same population.

Example 10.12: We revisit here the uncoated carbon data of Example 10.2. The data sets are intriguing due to their non-conformance with any of the standard distribution properties. Here we shall re-examine the data for the 30mm length. A histogram of these 132 observations is shown in Figure 10.11.

The bimodal nature of the data distribution suggests that a mixture distribution might be appropriate. It is first sensible, though, to reflect on the physical nature of the experiment and provide some plausible reason for a mixture model. There are two relevant facts here. First, the handling of fibres of approximately 7 micrometres in diameter risks introducing damage which could hasten failure. Second, it is known that failure can be initiated from surface flaws or from voids or inclusions within the material. There is thus likely to be some

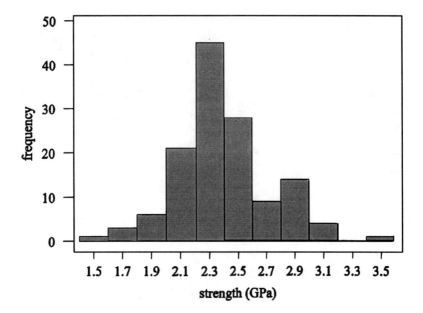

Figure 10.11. Strength distribution for 30mm uncoated carbon fibres

element of competing risks here, but in short fibres not all potential causes of failure may be present.

Suppose we consider a model which is part mixture, part competing risks. Let the higher-strength internal material flaws have reliability function $R_1(x)$ and the lower-strength surface flaws have reliability function $R_2(x)$. Internal flaws will be assumed to be present all the time and surface damage present a proportion p of the time. The model for the strength of a fibre is then given by

$$R(x) = pR_1(x)R_2(x) + (1 - p)R_1(x).$$

For both modes of failure a Weibull model is considered appropriate, so

$$R(x) = p\exp\left[-\left(\frac{x}{\alpha_1}\right)^{\beta_1}\right]\exp\left[-\left(\frac{x}{\alpha_2}\right)^{\beta_2}\right] + (1 - p)\exp\left[-\left(\frac{x}{\alpha_1}\right)^{\beta_1}\right]. \quad (10.7)$$

A point worth noting here is that while the Weibull distribution has the weakest-link property, (10.7) will not display the same characteristic.

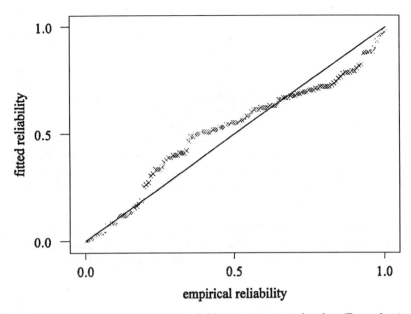

Figure 10.12. PP plot of the Weibull model for 30mm uncoated carbon (Example 10.12)

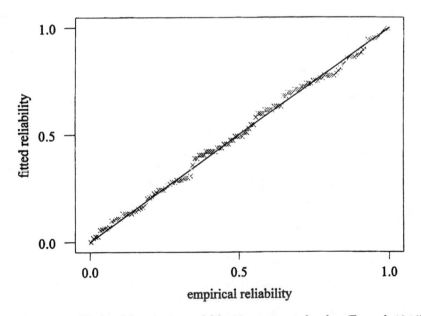

Figure 10.13. PP plot of the mixture model for 30mm uncoated carbon (Example 10.12)

Fitting (10.7) to the 30 mm uncoated carbon data yields the following parameter estimates: $\hat{p}$ = 0.617, $\hat{\alpha}_1$ = 2.78, $\hat{\beta}_1$ = 8.64, $\hat{\alpha}_2$ = 2.37, $\hat{\beta}_2$ = 17.87 and $\log L$ = −29.37. If a single Weibull distribution is fitted to the data, the estimated parameters are $\hat{\alpha}$ = 2.53 and $\hat{\beta}$ = 7.45 with maximized log likelihood −46.5. It is almost always the case that if the number of parameters in a model is increased then the likelihood is improved, but at the same time the standard errors of the estimates will be increased, as will any covariability between parameters. So it is important not to *over-fit* and to keep to what makes sense physically. Here physical justification for the model has been suggested and probability plots, of the form described in Section 4.7, are shown in Figure 10.12 for the single Weibull model and in Figure 10.13 for the fitted model (10.7). The log likelihood values reflect the degree of fit.

It is unfortunately the case that a likelihood ratio test is not straight-forward here. The reason is that if p is set to zero, the parameters α_2 and β_2 disappear. This problem is discussed by Davies (1987).

Useful Mathematical Techniques

A.1 Partial fractions

It is often useful to be able to reverse the following process:

$$\frac{a}{x+b} + \frac{c}{x+d} = \frac{a(x+d) + c(x+b)}{(x+b)(x+d)} = \frac{ex+f}{x^2 + (b+d)x + bd}.$$

Given an algebraic fraction where both the numerator and denominator are polynomials, the denominator is expressed in its simplest factorized form and each factor is the denominator of a term in the equivalent fraction sum. The following example illustrates how the numerators are determined.

Let

$$\frac{x+4}{x^2+3x+2} = \frac{x+4}{(x+1)(x+2)} = \frac{A}{x+1} + \frac{B}{x+2}.$$

By putting the right-hand side over a common denominator, as above, and then equating numerators,

$$x + 4 = A(x + 2) + B(x + 1).$$

Matching the terms in x and the constant terms yields

$$x = (A + B)x \quad \text{and} \quad 4 = 2A + B.$$

Hence $A = 3$ and $B = -2$, that is,

$$\frac{x+4}{x^2+3x+2} = \frac{3}{x+1} - \frac{2}{x+2}.$$

This rewriting of an algebraic fraction of this type is known as *partial fractions.*

It is important to note two general principles. First, the numerator of each fraction must be of a lower order than the denominator. For example, the following division would be necessary,

$$\frac{x^2 + 4}{x^2 + 3x + 2} = 1 + \frac{-3x + 2}{x^2 + 3x + 2},$$

before converting the second term by partial fractions. Second, the numerators of the partial fractions must be expressed initially as one order less than the denominator. For example,

$$\frac{x + 4}{x^3 - x^2 + 2x - 2} = \frac{x + 4}{(x^2 + 2)(x - 1)} = \frac{Ax + B}{x^2 + 2} + \frac{C}{x - 1}.$$

Equating numerators,

$$(Ax + B)(x - 1) + C(x^2 + 2) = x + 4.$$

Therefore,

$$A + C = 0,$$
$$-A + B = 1,$$
$$-B + 2C = 4.$$

So

$$A = -5/3, \ B = -2/3, \ C = 5/3.$$

A.2   Series

Algebraic sums occur in a variety of situations and may have a finite or infinite number of terms. Such sums may be known as *series,* and one of special interest is the *geometric series* which takes the form

$$a + ar + ar^2 + ar^3 + ar^4 + \ldots.$$

This series is said to have first term a and common ratio r. When the absolute value of r is less than 1, the terms in the series reduce in size rapidly and the sum of the series to an infinite number of terms may be evaluated.

Let the sum to n terms be S_n. Then

$$S_n - rS_n = a + ar + ar^2 + \ldots + ar^{n-1} - (ar + ar^2 + ar^3 + \ldots + ar^n).$$

So

$$S_n(1 - r) = a - ar^n,$$
$$S_n = \frac{a(1 - r^n)}{(1 - r)}.$$

For $|r| < 1$, r^n tends to zero as n tends to infinity, so

$$S_\infty = \frac{a}{1 - r}.$$

A.3   Taylor expansions

Algebraic expressions which prove problematic may be more tractable when expressed in an approximately equivalent form. Typically, a function $f(x)$ may be approximated by a polynomial series $a_0 + a_1 x + a_1 x^2 + \ldots$. Determination of suitable coefficients $\{a_i\}$ is done via *Taylor's theorem*. The justification for the following formula can be found in any intermediate mathematics text.

$$f(x - k) = f(k) + f'(k)(x - k) + \frac{f''(k)}{2!}(x - k)^2 + \ldots + \frac{f^{(n)}(k)}{n!}(x - k)^n + \ldots,$$

where $f^{(n)}(k)$ is the nth derivative of $f(x)$ with respect to x evaluated at $x = k$.

For practical purposes the series needs to be truncated after a modest number of terms. The approximation therefore only works adequately if the terms neglected are very small. Applications in this text use the special case $k = 0$, which is often referred to as the *Maclaurin expansion*. A very useful example of a Maclaurin expansion is that for $f(x) = e^x$. Here all $f^{(n)}(x) = e^x$, hence all $f^{(n)}(0) = 1$. So

$$e^x = 1 + x + \frac{x^2}{2!} + \frac{x^3}{3!} + \ldots + \frac{x^n}{n!} + \ldots .$$

Taylor expansions will yield finite series where derivatives of $f(x)$ eventually become zero. This will only occur when $f(x)$ is itself a polynomial. A special case of interest is the *binomial expansion*,

$$(a + b)^n = a^n + na^{n-1}b + \frac{n(n-1)}{2!}a^{n-2}b^2 + \ldots + nab^{n-1}b^n .$$

The example most found in reliability analysis is

$$(1 + x)^n = 1 + nx + \frac{n(n-1)}{2!}x^2 + \frac{n(n-1)(n-2)}{3!}x^3 + \ldots .$$

The value of x is often very small so a suitable approximation may be given by just the first two terms, that is,

$$(1 + x)^n \approx 1 + nx .$$

A.4   Newton–Raphson iteration

Many algebraic equations cannot be solved simply by rearranging the equation to make the term of interest the subject. A straightforward example is the following:

$$2 = x + \log x .$$

An iterative method of solution takes an initial trial solution $x = x_0$ and uses how well or otherwise the value fits the equation to generate a new, better value for x. The process continues until a value for x which satisfies the equation sufficiently closely is found. There are a number of possible iterative approaches, but the Newton–Raphson method is simple and has stood the test of time.

We start with an equation of the form $g(x) = 0$. For the example above, this means $x + \log x - 2 = 0$. The initial trial value x_0 is an informed guess in general. The next improved value of x, x_1, is generated by the formula

$$x_{n+1} = x_n - \frac{g(x_n)}{g'(x_n)}.$$

The process is repeated until x_{n+1} is approximately the same as x_n, to whatever tolerance is considered suitable.

The Newton–Raphson formula is in fact an adaptation of a Taylor expansion to two terms.

A.5   Numerical integration

Evaluation of definite integrals by numerical means, rather than by calculus, occurs when the integrand is a function which cannot arise as the derivative of some other function or where calculus yields a potentially infinite number of terms to evaluate. Examples arising in statistics include the normal and gamma probability density functions. Consider the standard normal distribution, which is the basis of normal tables, where

$$f(z) = \frac{1}{\sqrt{2\pi}} \exp\left(-\frac{z^2}{2}\right).$$

The probability that z lies between 0 and 2 is given by $\int_0^2 f(z) dz$.

However, there is no function which, when differentiated with respect to z, yields $\exp(-z^2/2)$. It is said that the integral of $f(z)$ does not have a *closed-form solution*.

A practical interpretation and use of integration is the calculation of areas between $f(z)$ and the z axis. Numerical methods for calculating such areas are based on subdividing the area into thin strips parallel to the $f(z)$ axis, approximating the area of each strip and then summing over all strips. The simplest approximation treats each strip as a trapezium, width $(z_{i+1} - z_i)$, parallel sides $f(z_i)$ and $f(z_{i+1})$, as in Figure A.1. Then

$$\int_{z_0}^{z_n} f(z) dz \approx \sum_{i=0}^{n-1} (z_{i+1} - z_i) \frac{f(z_i) + f(z_{i+1})}{2}.$$

The thinner the strips, that is, the closer z_{i+1} is to z_i, and consequently the larger the value of n, the more accurate the result.

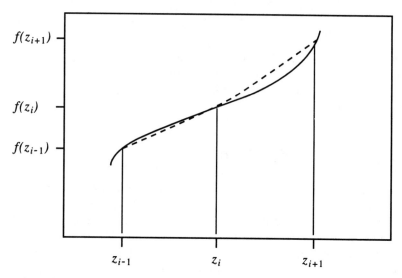

Figure A.1. Area under $y = f(z)$ divided into strips

More sophisticated methods attempt to model the top edge of a strip as a curve rather than a straight line. A popular method is *Simpson's rule*, which uses a quadratic approximation. It is convenient to make all the area strips the same width, so each $z_{i+1} - z_i$ will be denoted h. Simpson's rule requires that there is an even number of strips, in other words n is even, and results in

$$\int_{z_0}^{z_n} f(z)\mathrm{d}z \approx \frac{h}{3}[f(z_0) + 4f(z_1) + 2f(z_2) + 4f(z_3) + 2f(z_4) + \ldots$$

$$+ 4f(z_{n-1}) + f(z_n)].$$

A.6   Matrix algebra

Matrices are commonly first met as a shorthand way of presenting simultaneous equations. For example,

$$6 = 3x + y$$

$$5 = -x + 2y$$

may be described as $\mathbf{d} = \mathbf{M\theta}$, that is,

$$\begin{bmatrix} 6 \\ 5 \end{bmatrix} = \begin{bmatrix} 3 & 1 \\ -1 & 2 \end{bmatrix} \begin{bmatrix} x \\ y \end{bmatrix},$$

where $\mathbf{d}$ and $\mathbf{\theta}$ are *vectors* and $\mathbf{M}$ is a 2×2 matrix (with 2 rows and 2 columns). A column vector is in effect an $n \times 1$ matrix and a row vector a $1 \times n$ matrix.

Rows and columns of a matrix may be interchanged on occasion and the result, $\mathbf{M^T}$, is called the *transpose* of the matrix $\mathbf{M}$. So if

$$\mathbf{M} = \begin{bmatrix} 3 & 1 \\ -1 & 2 \end{bmatrix}, \quad \mathbf{M^T} = \begin{bmatrix} 3 & -1 \\ 1 & 2 \end{bmatrix}.$$

Solving equations of the form $\mathbf{d} = \mathbf{M\theta}$ for θ requires the *inverse* of the matrix $\mathbf{M}$, denoted $\mathbf{M^{-1}}$, so that

$$\mathbf{M^{-1}d} = \mathbf{\theta}.$$

Matrices in this context are by definition square, and for the 2×2 case a simple formula exists for the inverse. If

$$\mathbf{M} = \begin{bmatrix} a & b \\ c & d \end{bmatrix},$$

then

$$\mathbf{M^{-1}} = \frac{1}{ad - bc} \begin{bmatrix} d & -b \\ -c & a \end{bmatrix}.$$

The denominator in this expression is known as the *determinant* of the matrix.

For matrices of higher dimension, computing assistance is desirable for finding inverses. Statistical packages will invert matrices with numerical entries and computer algebra packages will invert matrices with algebraic entries.

A.7   The principle of least squares

In simple linear regression a straight line is fitted to a set of points (x, y). The criterion upon which a line is fitted in some 'best' sense is commonly that of *least squares*. The values of x recorded are realizations of an explanatory variable, something which is considered to influence the corresponding y observation. The fitted straight line serves to determine what the y reading would be, on average, for a given x reading. So for every observed x_i there is an observed y_i and a predicted y, $\hat{y}_i$, obtained from the fitted linear relationship $y = mx + c$. Estimation of the constants m and c depends on the fitting criterion.

The least squares principle is that the sum of the squared deviations of the distances between the points and the line, in the y direction, should be minimized. Define

$$R^2 = \sum_{i=1}^{n} [y_i - \hat{y}_i]^2,$$

where n is the number of observations. If $\hat{y}$ is of linear form,

$$R^2 = \sum_{i=1}^{n} [y_i - (mx_i + c)]^2.$$

Least squares estimates of m and c are obtained by differentiating R^2 with respect to m and with respect to c and setting each derivative to zero:

$$\frac{\partial R^2}{\partial c} = -2 \sum_{i=1}^{n} [y_i - (mx_i + c)] = 0,$$

$$\frac{\partial R^2}{\partial m} = -2 \sum_{i=1}^{n} x_i[y_i - (mx_i + c)] = 0.$$

This leads to the matrix equation

$$\begin{bmatrix} n & \sum x_i \\ \sum x_i & \sum x_i^2 \end{bmatrix} \begin{bmatrix} c \\ m \end{bmatrix} = \begin{bmatrix} \sum y_i \\ \sum x_i y_i \end{bmatrix}.$$

Solving by the methods shown in Section A.5 yields

$$\hat{c} = \frac{\sum x_i^2 \sum y_i - \sum x_i \sum x_i y_i}{n \sum x_i^2 - (\sum x_i)^2},$$

$$\hat{m} = \frac{n \sum x_i y_i - \sum x_i \sum y_i}{n \sum x_i^2 - (\sum x_i)^2}.$$

It can be shown that $\bar{y} = \hat{m}\bar{x} + \hat{c}$, where $\bar{y}$ and $\bar{x}$ are the means respectively of the $\{y_i\}$ and $\{x_i\}$, so that, given an estimate of one parameter, the other may be estimated by this simple relationship.

References

Ansell, J.I. and Phillips, M.J. (1994) *Practical Methods for Reliability Data Analysis*, Oxford University Press, New York.

Ansell, J.I. and Walls, L.A. (1992) Dependency analysis in reliability studies. *IMA J. Math. Appl. Bus. Indust.*, **3**, 333–348.

Ascher, H. and Feingold, H. (1984) *Repairable Systems Reliability*, Marcel Dekker, New York.

Barbour, A.D. (1981) Brownian motion and a sharply curved boundary. *Adv. Appl. Probab.*, **13**, 736–750.

Barlow, R.E. and Proschan, F. (1996) *Mathematical Theory of Reliability*, SIAM, Philadelphia.

Bates, G.E. (1955) Joint distribution of time intervals for the occurrence of successive incidents in a generalised Polya scheme. *Ann. Math. Statist.*, **26**, 705–720.

Chatfield, C. (1983) *Statistics for Technology*, 3rd edition, Chapman & Hall, London.

Coles, S.C. (1989) On goodness-of-fit tests for the two parameter Weibull distribution derived from the stabilized probability plot. *Biometrika*, **76**, 593–598.

Cox, D.R. (1962) *Renewal Theory*, Methuen, London.

Cox, D.R. (1972) Regression models and life tables (with discussion). *J. Roy. Statist. Soc. B*, **34**, 187–220.

Cox, D.R. and Lewis, P.A.W. (1966) *The Statistical Analysis of Series of Events*, Methuen, London.

Cox, D.R. and Oakes, D. (1984) *Analysis of Survival Data*, Chapman & Hall, London.

Crowder, M.J., Kimber, A.C., Smith, R.L. and Sweeting, T.J. (1991) *Statistical Analysis of Reliability Data*, Chapman & Hall, London.

Cunnane, C. (1978), Unbiased plotting positions – a review. *J. Hydrology*, **37**, 205–222.

D'Agostino, R.B. and Stephens, M.A. (1986) *Goodness-of-fit Techniques*, Marcel Dekker, New York.

Daniels, H.E. (1945) The statistical theory of the strength of bundles of threads. I. *Proc. Roy. Soc. London A*, **183**, 404–435.

Davies, R.B. (1987) Hypothesis testing when a nuisance parameter is present only under the alternative. *Biometrika*, **74**, 33–43.

Davis, D.J. (1952) An analysis of some failure data. *J. Amer. Statist. Assoc.*, **47**, 113–150.

Duane, J.T. (1964) Learning curve approach to reliability monitoring. *IEEE Trans. Aerospace*, **2**, 563–566.

Dumonceaux, R. and Antle, C.E. (1973) Discrimination between the lognormal and Weibull distributions. *Technometrics*, **15**, 923–926.

Fleming, K.N. (1975) A reliability model for common mode failures in redundant safety systems. General Atomic Report No. GA-A13284.

Greenwood, M. (1926) The natural duration of cancer. *Reports on Public Health and Medical Subjects*, 33, 1–26. HMSO, London.

Hinds, P.A. (1996) Reliability analysis techniques for equipment management. MSc dissertation, City University, London.

Hines, W.M. and Montgomery, D.C. (1990) *Probability and Statistics in Engineering and Management Science*, 3rd edition, Wiley, New York.

Jelinski, Z. and Moranda, P.B. (1972) Software reliability research, in *Statistical Computer Performance Evaluation* (ed. W. Freiberger), Academic Press, London.

Kalbfleisch, J. and Prentice, R.L. (1980) *The Statistical Analysis of Failure Time Data*, Wiley, New York.

Kaplan, E.L. and Meier, P. (1958) Nonparametric estimation from incomplete observations. *J. Amer. Statist. Assoc.*, **53**, 457–481.

Kapur, K.C. and Lamberson, L.R. (1977) *Reliability in Engineering Design*, Wiley, New York.

Kimber, A.C. (1985) Tests for the exponential, Weibull and Gumbel distributions based on the stabilized probability plot. *Biometrika*, **72**, 661–663.

Klaasen, K.B. and van Peppen, J.C.L. (1989) *System Reliability: Concepts and Applications*, Edward Arnold, London.

Lawless, J.F. (1982) *Statistical Models and Methods for Lifetime Data*, Wiley, New York.

Leadbetter, M.R., Lindgren, G. and Rootzen, H. (1983) *Extremes and Related Properties of Random Sequences and Processes*, Springer-Verlag, New York.

Leitch, R.D. (1995) *Reliability Analysis for Engineers*, Oxford University Press, New York.

Lieberman, G.J. and Ross, S.M. (1971) Confidence intervals for independent exponential series systems. *J. Amer. Statist. Assoc.*, **66**, 837–840.

Lieblein, J. and Zelen, M. (1956) Statistical investigation of the fatigue life of deep groove ball bearings. *J. Res. Nat. Bur. Standards*, **57**, 273–316.

Lindsey, J.K. (1997) Parametric multiplicative intensities models fitted to bus motor failure data. *J. Appl. Statist.*, **46**, 245–252.

Lipow, M. and Riley, J. (1960) *Tables of Upper Confidence Limits on Failure Probability of 1, 2 and 3 Component Serial Systems, Vols 1 and 2*. US Dept. of Commerce AD-609-100 and AD-636-718, Clearinghouse, Washington, DC.

Lloyd, D.K. and Lipow, M. (1962) *Reliability: Management Methods and Mathematics*, Prentice Hall, London.

Mann, N.R. and Fertig, K.W. (1973) Tables for obtaining confidence bounds and tolerance bounds based on best linear invariant estimates of parameters of the extreme value distribution. *Technometrics*, **16**, 335–346.

Mann, N.R., Schafer, R.E. and Singpurwalla, N.D. (1974) *Methods for Statistical Analysis of Reliability and Lifetime Data*, Wiley, New York.

McCartney, L.N. and Smith, R.L. (1983) Statistical theory of the strength of fiber bundles. *J. Appl. Mech.*, **50**, 601–608.

Metcalfe, A.V. (1994) *Statistics in Engineering*, Chapman & Hall, London.

Michael, J.R. (1983) The stabilized probability plot. *Biometrika*, **70**, 11–17.

Mudholkar, G.S., Srivastava, D.K. and Freimer, M. (1995) The exponentiated Weibull family: a reanalysis of the bus-motor-failure data. *Technometrics*, **37**, 436–445.

Neave, H.R. (1985) *Elementary Statistics Tables*, George Allen & Unwin, London.

Nelson, W.B. (1982) *Applied Life Data Analysis*, Wiley, New York.

Nelson, W.B. (1988) Graphical analysis of system repair data. *J. Quality Technol.*, **20**(1), 24–25.

Nelson, W.B. (1993) *Accelerated Life Testing*, Wiley, New York.

Pearson, E.S. and Hartley, H.O. (1972) *Biometrika Tables for Statisticians, Vol. II*, Cambridge University Press, Cambridge.

Press, W.H., Flannery, B.P., Teukolsky, S.A. and Vetterling, W.T. (1992) *Numerical Recipes*, 2nd edition, Cambridge University Press, Cambridge.

Proschan, F. (1963) Theoretical explanation of observed decreasing failure rate. *Technometrics*, **5**, 375–383.

Rosenblatt, J.R. (1963) Confidence limits for the reliability of complex systems, in *Statistical Theory of Reliability* (ed. M. Zelen), University of Wisconsin Press, Madison.

Shapiro, S.S. and Wilk, M.B. (1965) An analysis of variance of test for normality (complete samples). *Biometrika*, **52**, 591.

Smith, D.J. (1993) *Reliability, Maintainability and Risk*, 4th edition, Butterworth-Heinemann, Oxford.

Smith, R.L. (1991) Weibull regression models for reliability data, *Reliability Engrg. System Safety*, **34**, 35–57.

Soper, S. and Wolstenholme, L.C. (1993) Morse taxonomy of architectures. Technical Report Morse/Transmitton/SRKS/4/V1 for Department of Trade and Industry report SCSATP-IED4/1/9001.

Stephens, M.A. (1977) Goodness of fit for the extreme value distribution. *Biometrika*, **62**, 467–475.

Stuart, A. and Ord, J.K. (1994) *Kendall's Advanced Theory of Statistics, Volume 1*, 6th edition, Edward Arnold, London.

Thompson, W.A. Jr (1988) *Point Process Models with Applications to Safety and Reliability*, Chapman & Hall, London.

Watson, A.S. and Smith, R.L. (1985) An examination of statistical theories for fibrous materials in the light of experimental data. *J. Materials Sci.*, **20**, 3260–3270.

Winterbottom, A. (1974) Lower confidence limits for series system reliability from binomial subsystem test data. *J. Amer. Statist. Assoc.*, **69**, 782–788.

Winterbottom, A. (1980) Asymptotic expansions to improve large sample confidence intervals for system reliability. *Biometrika*, **67**, 351–357.

Winterbottom, A. (1984) The interval estimation of system reliability from component test data. *Oper. Res.*, 32, 628–640.

Wolstenholme, L.C. (1991) A dependent bundles model for estimating stress concentrations in fibre-matrix composites. *J. Materials Sci.*, **26**, 4599–4614.

Wolstenholme, L.C. (1995) A non-parametric test of the weakest-link property. *Technometrics*, **37**, 169–175.

Wolstenholme, L.C. (1996) An alternative to the Weibull distribution. *Commun. Statist. Simulation Comput.*, **25**(1), 119–137.

Index